Inclusive Tourism Futures

THE FUTURE OF TOURISM

Series Editors: Ian Yeoman, Victoria University of Wellington, New Zealand and **Una McMahon-Beattie**, *Ulster University, Northern Ireland, UK.*

Some would say that the only certainties are birth and death; everything else that happens in between is uncertain. Uncertainty stems from risk, a lack of understanding or a lack of familiarity. Whether it is political instability, autonomous transport, hypersonic travel or peak oil, the future of tourism is full of uncertainty but it can be explained or imagined through trend analysis, economic forecasting or scenario planning.

This new book series, The Future of Tourism, sets out to address the challenges and unexplained futures of tourism, events and hospitality. By addressing the big questions of change, examining new theories and frameworks or critical issues pertaining to research or industry, the series will stretch your understanding and generate dialogue about the future. By adopting a multidisciplinary perspective, be it through science fiction or computer-generated equilibrium modelling of tourism economies, the series will explain and structure the future - to help researchers, managers and students understand how futures could occur. The series welcomes proposals on emerging trends and critical issues across the tourism industry and research. All proposals must emphasise the future and be embedded in research.

All books in this series are externally peer-reviewed.

Full details of all the books in this series and of all our other publications can be found on http://www.channelviewpublications.com, or by writing to Channel View Publications, St Nicholas House, 31–34 High Street, BristolBS1 2AW, UK.

THE FUTURE OF TOURISM: 5

Inclusive Tourism Futures

Edited by
**Anu Harju-Myllyaho and
Salla Jutila**

CHANNEL VIEW PUBLICATIONS
Bristol • Blue Ridge Summit

DOI https://doi.org/10.21832/HARJU6874

Library of Congress Cataloging in Publication Data
A catalog record for this book is available from the Library of Congress.
Names: Harju-Myllyaho, Anu, 1982- editor. | Jutila, Salla, 1983- editor.
Title: Inclusive Tourism Futures/Edited by Anu Harju-Myllyaho and Salla Jutila. Description: Bristol, UK ; Blue Ridge Summit, PA : Channel View Publications, 2021. | Series: The Future of Tourism: 5 | Includes bibliographical references and index. | Summary: "This book combines studies of inclusivity in tourism with a future lens and provides timely insights into current research and discussions on social inclusion. It considers a future that can be welcoming of different ways of being, doing and knowing to empower all participants in the planning and development of tourism and hospitality"—Provided by publisher.
Identifiers: LCCN 2020056194 (print) | LCCN 2020056195 (ebook) | ISBN 9781845416867 (paperback) | ISBN 9781845416874 (hardback) | ISBN 9781845416881 (pdf) | ISBN 9781845416898 (epub) | ISBN 9781845416904 (kindle edition) Subjects: LCSH: Tourism—Social aspects. | Tourism—Planning. | Social integration. Classification: LCC G156.5.S63 I64 2021 (print) | LCC G156.5.S63 (ebook) | DDC 338.49104—dc23 LC record available at https://lccn.loc.gov/2020056194
LC ebook record available at https://lccn.loc.gov/2020056195

British Library Cataloguing in Publication Data
A catalogue entry for this book is available from the British Library.

ISBN-13: 978-1-84541-687-4 (hbk)
ISBN-13: 978-1-84541-686-7 (pbk)

Channel View Publications
UK: St Nicholas House, 31–34 High Street, Bristol, BS1 2AW, UK.
USA: NBN, Blue Ridge Summit, PA, USA.

Website: www.channelviewpublications.com
Twitter: Channel_View
Facebook: https://www.facebook.com/channelviewpublications
Blog: www.channelviewpublications.wordpress.com

Copyright © 2021 Anu Harju-Myllyaho, Salla Jutila and the authors of individual chapters.

All rights reserved. No part of this work may be reproduced in any form or by any means without permission in writing from the publisher.

The policy of Multilingual Matters/Channel View Publications is to use papers that are natural, renewable and recyclable products, made from wood grown in sustainable forests. In the manufacturing process of our books, and to further support our policy, preference is given to printers that have FSC and PEFC Chain of Custody certification. The FSC and/or PEFC logos will appear on those books where full certification has been granted to the printer concerned.

Typeset by SAN Publishing Services.

Contents

Contributors		vii
Preface		xi
Introduction *Anu Harju-Myllyaho and Salla Jutila*		1

Part 1: Actors

1 Hosts and Guests in Participatory Development 15
   Emily Höckert, Outi Kugapi and Monika Lüthje

2 Resourcing the Self: Forms of Capital and Stratification
   in the Platform Hospitality Community 33
   Andreja Trdina, Salla Jutila and Maja Turnšek

Part 2: Methods

3 Inclusion in Tourism Strategies: Setting the Stage
   for Inclusive Tourism Development in Tourism Destinations 59
   Anu Harju-Myllyaho and Salla Jutila

4 Where is 'The Poor' in Pro-poor VCA? – A Review
   of Applying Pro-poor VCA in a Coastal Tourism Destination
   in the Northeast of Brazil 80
   Theres Winter, Seonyoung Kim and Nicola Palmer

Part 3: Practices

5 Information about Tourism Destinations' Accessibility
   in Tourism Online Platforms: Is it Useful for People
   with Diverse Abilities? 107
   *Asunción Fernández-Villarán, Mónica Erice,
   Nagore Espinosa, Ana Goytia, Aurora Madariaga
   and Ainara Rodríguez*

6	Current Perspectives on Social Inclusion in Tourism in Finnish Lapland *Sari Nisula, Marlene Kohllechner-Autto and Krista Skantz*	122
7	Conclusion *Anu Harju-Myllyaho and Salla Jutila*	144
Index		154

Contributors

Editors

Anu Harju-Myllyaho works as Head of Expertise Group, Responsibility in Business and Services, at the Multidimensional Tourism Institute, Lapland University of Applied Sciences, Finland. She has a master's degrees in hospitality management (2012) and social sciences (2018). She is currently working on her PhD at the University of Lapland. Her research interests concern inclusion in tourism, tourism employment and futures studies.

Salla Jutila is a university teacher and PhD student at the University of Lapland Multidimensional Tourism Institute. Her research and teaching interests include inclusion in tourism, sharing economy in tourism, accessible tourism, sustainable tourism planning and tourism foresight. She has authored and co-authored publications in Finnish tourism journals as well as several book chapters. Jutila has been involved in regional, national and international tourism development projects as project manager, expert and project planner.

Authors

Emily Höckert is a Postdoctoral Researcher at the University of Lapland Multidimensional Tourism Institute. She wrote *Negotiating Hospitality* (Routledge, 2018) and co-authored *Disruptive Tourism and Its Untidy Guests* (Palgrave Macmillan, 2014), both of which focus on exploring relational ethics in tourism. Today, she is a proud member of the research and development teams Culturally Sensitive Tourism in the Arctic (ARCTISEN) and Envisioning Proximity Tourism with New Materialism (Academy of Finland, 324493).

Outi Kugapi is a Doctoral Candidate in tourism research at University of Lapland Multidimensional Tourism Institute. She knits and purls together her PhD, warm mittens and two projects concerning cultural sensitivity (ARCTISEN, Northern Periphery and Arctic Programme) and handicraft tourism (Handmade in Lapland, European Social Fund (ESF)). Her research interests are tourist experiences in craft tourism and she co-authored the article 'Affective entanglements with travelling mittens' (*Tourism Geographies*, 2020) as well as a chapter about crafts as employment and cultural experience service in *Tourism Employment in Nordic Countries: Trends, Practices, and Opportunities* (Palgrave Macmillan, 2020). She has also co-authored chapters on Indigenous tourism in several books.

Monika Lüthje is a Senior Researcher in Tourism Research at the University of Lapland. Her current research interests are cultural sensitivity and practice-based approaches as ways to develop Arctic tourism more responsibly towards local cultures. She is especially interested in Sámi tourism and currently works as the responsible leader of the project Culturally Sensitive Tourism in the Arctic (ARCTISEN), funded by the EU's Northern Periphery and Arctic Programme.

Andreja Trdina is an Assistant Professor at the Faculty of Tourism, University of Maribor, Slovenia. She has an academic background in media studies and has, in her research, focused on popular culture and media, sociology of taste, class and distinction with special regard to contemporary material/consumer culture. She is currently mostly dealing with research on mediatisation of tourism, travel as social and cultural practice, and issues of mobility and belonging. Among others, her contributions have been published by Routledge and in the journals *Slavic Review*, *Javnost – The Public*, *Tourism: An International Interdisciplinary Journal* and *Comedy Studies*.

Maja Turnšek is Associate Professor and Vice Dean for Research at the Faculty of Tourism, University of Maribor, Slovenia. Her background is in media and communication studies. She lectures on communication, psychology and marketing in tourism. Her main research interests cover platforms in tourism, the political economy of new media, sharing economy, travel-related platform work, storytelling, humour and experience design in tourism.

Theres Winter completed her PhD in 2018 and is Associate Lecturer at Sheffield Hallam University, UK and Research Associate at the University for Sustainable Development in Eberswalde, Germany. She researches tourism, poverty and inequality, and visual methodologies.

Seonyoung Kim is a Principal Lecturer in Tourism Management at Sheffield Business School, Sheffield Hallam University, UK. She has research interests in tourism governance, tourism policy and planning, urban tourism, sustainable development and accessible tourism.

Nicola Palmer has worked in tourism for 25 years as an educator, consultant and researcher. Her interests align with tourism development studies, transitional economies, ecotourism, governance and cultural change. She has developed networks across the public sector and in the visitor economy. She has supervised over 17 PhD students to completion and has a wealth of experience in teaching research methods in social science at postgraduate level. Her publications focus on people, place and space. Nicola is particularly concerned with issues of equity, participation, identity and socioeconomic transformation and she is an active volunteer in the fields of heritage and education literacy.

Asunción Fernández-Villarán currently works as a Lecturer with PhD at the University of Deusto, Bilbao, Spain. She teaches BA courses in Tourism and MA courses in Leisure Project Management and Congress, Events and Fairs Management. She leads the research group Tourism at the University of Deusto. Her scope of research focuses on tourism, advanced measurement systems (big data) and smart solutions, tourism for all, tourism and Sustainable Development Goals (SDG), and tourism intermediation. She is the author of a number of articles dealing with tourism.

Mónica Erice is a BA in Political Science and Sociology from the Complutense University of Madrid and MA in European Integration: the European Union from the University of the Basque Country. She is currently Lecturer at the University of Deusto (Bilbao, Spain). She teaches BA courses in Tourism and is coordinator of the Department of External Practical Training attached to qualifications. Her publications mainly focus on the analysis and diagnosis of tourism on a regional level, cultural tourism, tourism for all and smart tourism. She forms part of the Human Leisure and Development research team recognised by the Basque Government.

Nagore Espinosa is CEO of the research and consultancy company in2destination, Vice President of the INRouTe International Network, Expert Consultant for the United Nations World Tourism Organization, for the United Nations International Trade Center and for the Inter-American Tourism Bank and a Lecturer at the University of Deusto. With more than 15 years of experience, she advises on competitive, sustainable and inclusive tourism destinations strategy and tourism measurement to national, regional and local governments in Spain, Colombia, Brazil, Myanmar, Kazakhstan and Cape Verde, among others. She holds a doctorate in Business Competitiveness and Economic Development and has written numerous academic and informative publications.

Ana Goytia is a Full Professor at the Faculty of Social and Human Sciences (Department of Tourism) of the University of Deusto. Her work focuses on the study of tourism experience from a psycho-sociological perspective. Her areas of knowledge are located in the following fields: the psycho-sociology of tourism demand; tourism and lifestyles; tourism for all and tourism demand needs; tourism and sustainable development goals (UN SDGs); policies and strategic planning for sustainable and inclusive tourism development; (UNESCO codes: 630200 Experimental Sociology, 531290 Tourism, 530802 Consumer Behavior).

Aurora Madariaga is a Professor and Researcher with a six-year research period and another six-year period of social transference at the Institute of Leisure Studies of the Faculty of Social and Human Sciences, University of Deusto. She is the main researcher of the official research group (type A ref. IT984-16) on leisure as a factor of human development. Her lines of research are related to inclusion in leisure, people with disabilities, leisure education and the third sector. She directs the Chair of Leisure and Disability and promotes the Inclusive Leisure Manifesto.

Sari Nisula has an MA from the University of Lapland. She is a project planner and lecturer, and has worked on a variety of projects concerning aspects of tourism, sustainability and sparsely populated areas.

Marlene Kohllechner-Autto is a graduate of the Management Center Innsbruck, Austria, and holds an MBA from Lapland University of Applied Sciences. Working as Project Manager at Lapland UAS, her focal point has mainly been on international projects centred around tourism, sparsely populated areas and social economy. Having previously worked in the tourism industry in Lapland, the potential impact of tourism has on sparsely populated areas is of great interest to her.

Krista Skantz has a bachelor's degree in Hospitality Management from the Lapland University of Applied Sciences. From the very beginning, the topic of equality inspired and steered her through her studies of tourism. Krista has taken part in developing projects as a trainee and worked on issues of social inclusion, social entrepreneurship in sparsely populated areas and accessible hospitality. She has also been working in customer service in different fields and as a personal assistant for people with disabilities. The above-mentioned social aspects still continue to inspire her and will surely find a place in her future career.

Preface

We met eight years ago when we started working at the Multidimensional Tourism Institute (MTI) during a renovation of the campus on which MTI was located at the time. We started work on the exact same day and while other staff were still on their summer leave, we supported each other in the new situation. We had both just graduated from master's programmes and were working as project planners on projects related to tourism foresight. In retrospect, everything that happened after that has led to this point, where we are now: working on our PhDs and this book related to a topic that has become important to us.

Working on the project Foresight as a Competitive Advantage for Tourism in Lapland and coordinating the national tourism foresight network were inspirational starting points for the theme of this book. Through these projects, we began to work with accessible tourism and foresight-related studies and projects in Lapland. We soon discovered that the subject touches more varied groups of people than one might think. We stayed with the concept of accessible hospitality for a while and this helped to clarify the issue of multidimensional accessibility and the challenges and possibilities that might emerge concerning the inclusion of different groups.

Over the years, we have worked with inclusion – whether on a project, during joint writing or while teaching. One idea or realisation led to another and that is how we built our view on what inclusive tourism is and how it should be developed. Our view has gained in depth during the writing process for this book as well.

Quite a few thanks are in order. First of all, thank you to the authors who have contributed to this book and shared their knowhow. Many thanks go to our colleagues at MTI – you inspire us to do our best. Thank you, Sanna Kyyrä, you are incredible. Johan Edelheim, thank you for creating the possibilities that led to this project. Special thanks to Sarah Williams for her advice and patience.

To Salla, thank you for your companionship.

To Anu, thank you for sharing so many experiences.

Introduction

Anu Harju-Myllyaho and Salla Jutila

Inclusion. It is a familiar and extensively used term, but what does it actually mean? An accurate and explicit definition may not be possible, as definitions are always dependent on the field and context in which they are made. Instead, it is possible to open up various aspects and dimensions of the term in a particular context. The principles of inclusion appeared for the first time in public discussion when the United Nations proclaimed the year 1981 as the International Year of Disabled Persons (UNESDOC, 1982). Later, inclusion became a central topic in the fields of education, special education and health, while inclusive design has become a common term in architecture and design. Gradually, inclusion has spread into wider societal and political discussions. According to the online Cambridge Dictionary (2020), inclusion refers to the idea that everyone should be able to use the same facilities, take part in the same activities and enjoy the same experiences. Inclusion also means allowing many different types of people to do something and treating them fairly and equally. Isola *et al.* (2017) claim that inclusion relates to participation, representation and democracy, and consists of involvement, relatedness, belonging and togetherness.

In the context of tourism, inclusion has been studied from several different stakeholders' points of view. Probably the most examined viewpoint is that of disabled people, their rights and opportunities to participate in tourism (e.g. Buhalis & Darcy, 2011; Darcy *et al.*, 2020). Another group that has gained lots of attention is the ageing population, given the economic significance of this segment with diverse special needs (e.g. Nielsen, 2013; Zsarnoczky *et al.*, 2016). The inclusion of low-income families and other groups who need financial support to participate in tourism has been discussed, especially in the field of social tourism (e.g. McCabe, 2009; McCabe & Johnson, 2013; Minneart *et al.*, 2009). Other actors whose voices as tourists have been brought into the inclusive tourism discussion are rainbow tourists (e.g. Harju-Myllyaho, 2018), ethnic minorities (e.g. Klemm, 2002) and tourists of different religions (e.g. Battour *et al.*, 2018; Jafari & Scott, 2014). Along with the diversity of tourists, inclusive tourism emphasises the participation of local people (e.g. Höckert, 2015; Simmons, 1994). Local stakeholders who have provoked particular discussion are Indigenous peoples (e.g. Butler & Hinch, 2007; Nielsen & Wilson, 2012; Zeppel, 2007) and ethnic minorities (e.g. Yang & Wall, 2009). An important group that has gained relatively little attention – but is relevant especially when keeping an eye on the future – is local children and young people in tourism destinations. Canosa *et al.* (2016) demonstrated and raised awareness concerning the lack of young people's voices, engagement and participation in tourism research and within host communities.

In the field of tourism, inclusion has often been approached through different forms of tourism with diverse emphases – such as accessible tourism, social tourism, participatory tourism, Indigenous tourism, tourism for all, community-based tourism and pro-poor tourism (see Haanpää *et al.*, 2018: 47). Each approach emphasises the viewpoints of different actors. Accessible tourism, social tourism and tourism for all point out the diversity of tourists and everyone's right to participate in tourism – people should be made welcome whatever their background or physical capabilities. Participatory tourism, Indigenous tourism, community-based tourism and pro-poor tourism, on the other hand, highlight the importance of local participation. Inclusive tourism covers both approaches – those that emphasise tourists' perspectives and those that focus on the perspectives of local people. Inclusive tourism thus encompasses the perspectives of all tourism stakeholders. As Biddulph and Scheyvens claim (2018: 584), inclusive tourism can be defined as tourism that engages different groups in the production and consumption of tourism and the sharing of its benefits.

Figure I.1 outlines the complex and multidimensional whole that forms inclusive tourism. Inclusive tourism is surrounded by the elements and foundations of which it is comprised. These elements are achieved

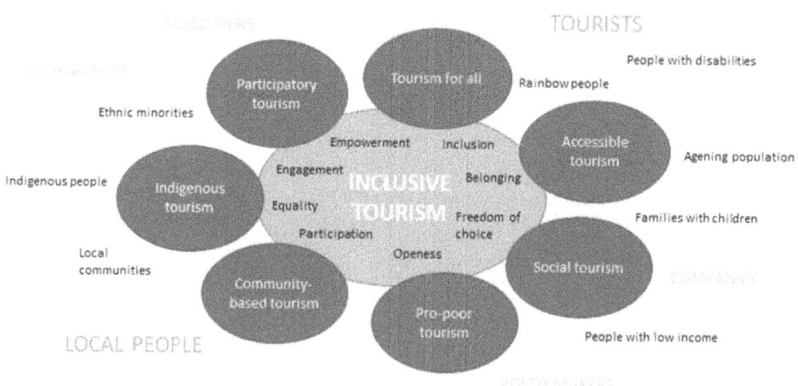

Figure I.1 Dimensions of inclusive tourism
Source: Authors

through different forms of tourism emphasising inclusion of a particular group of people. Different groups of people related to inclusive tourism can be described as target groups (in black letters on the edges of Figure I.1) or as positions from whose perspective inclusion can be approached (in grey capital letters in Figure I.1). Certainly, this diagram is not comprehensive. The forms of tourism and the groups of people in it are examples, but there are also other forms of tourism that support inclusion and other groups of people related to inclusive tourism.

Based on the different viewpoints and approaches presented above, it is justifiable to claim that inclusive tourism is a wide and multidimensional phenomenon. This is also what we discovered while compiling this book, the aim of which is to provide current viewpoints on tourism inclusion with a futures perspective. The idea is not to draw specific scenarios for inclusive tourism futures, but rather to provide timely insight into the discussion and lead the reader to consider the future possibilities based on current research and development streams. While various scholars have contributed to the discussion concerning inclusive tourism, especially in terms of different stakeholders and actors, there is still room for studies that take a futures perspective on inclusion. Our aim is to pay attention to inclusive tourism futures-making, which in this book is tied to methods and practices. How to understand and promote inclusive tourism development in academia and in the field are the questions of interest here. Thus, the purpose of this book is to compile a basis for further discussion as well as to understand the various tourism stakeholder groups, viewpoints, methods and practices that are important for supporting inclusive tourism. Practical examples can provide vistas from which to ponder best practices and different models, which can be contextualised and applied geographically and in different sectors in the field of tourism and hospitality. In this respect, this book does not provide an exhaustive picture of inclusive tourism, but rather offers a possibility to comprehend the vastness of the topic and to point out the importance of inclusion in producing ethical and sustainable tourism.

Tourism Inclusion in Light of Current Development Streams and Discussions

In recent decades, tourism has grown rapidly and has become an important part of the world economy. Its economic significance has been measured, although it is difficult to do so because of multiplicative effects and unregistered overnights, for instance. The ethos of growth has been the subject of extensive critical discourse. The negative consequences of tourism, such as over-tourism, environmental impact, and cultural and social appropriation, have become a part of daily tourism-related conversation. Tourism has, in a way, become a less fashionable form of consumerism; a well-educated world citizen might not want to take part in this kind of activity. At the very least, customers have become more demanding in terms of sustainability. This new form of ethical or responsible consumerism takes into consideration not only the environmental, but also the cultural and social aspects of tourism.

Although the current discussion concerning responsible tourism revolves around climate change and its environmental consequences, socioeconomic considerations have gained ground. None of the viewpoints on sustainable tourism exist in isolation. For instance, ecologically sustainable tourism demands a recognition of its social dimension. It is very important to continue discussions and to commit to action regarding socially responsible tourism and environmental sustainability. However, such engagement also causes conflicts. If we must restrict tourism or compensate for the emissions it causes through taxation, who then has the right to travel? Does anyone? At least from the viewpoint of mobility, everyone has the right – the human right – to movement (on which, see Chapter 1). At the same time, we ask: who is permitted to take part in planning tourism and consuming it? Who benefits?

Responsible tourism is part of a wider societal discussion that centres on our common fate and the fates of those who will follow us (see e.g. Rubin, 2004). It is important to give a voice to future generations and to consider different possibilities and paths that might be pursued to achieve mutually desired goals. Anticipating alternative futures allows us to make preparations, while recognising different perspectives on inclusive tourism makes it possible to consider future actors and what might be best for them.

Having said that, while editing this book, the world saw an unprecedented crisis that had a massive impact on tourism – the COVID-19 crisis has caused severe restrictions to mobility and has put tourism on hold, at least on an international level. Baum and Nguyen (2020) note that the COVID-19 crisis will impact freedom of movement as a right and, during the acute pandemic, it already has. It can be suspected that this crisis will not increase inclusive tourism. In the long term, anticipating the impact of the crisis is difficult. How will companies survive? Will governments open borders or will there be new waves of the virus? Will governments place more restrictions? These are questions that do not have clear answers at the time this book was written (see Baum & Nguyen, 2020; Hall *et al.*, 2020).

We do not specifically scrutinise the COVID-19 crisis or its impact on tourism in this book. Nevertheless, our aim is to understand how we can affect tourism inclusion and develop it further, while keeping in mind that occasionally we experience a black swan (e.g. Heinonen, 2013; Pramod 2020) – a phenomenon we cannot anticipate. Despite their randomness, black swans have a major impact on society as a whole. Even though the COVID-19 pandemic does not fit to all the characteristics of a black swan (see e.g. Inayatullah & Black, 2020), it took many of us by surprise. Heinonen (2013: 15) observes that people seek explanations for these events to make them seem less random and more predictable than they were. However, such events exceed all probabilities and that is why we must see them as complex uncertainties and future possibilities. (Heinonen, 2013: 20). Black swans, according to Heinonen (2013: 22), have a significant impact not only on our futures, but also on our worldviews; they do not fit in our common ways of thinking and how we see the world. In this sense, most people – especially in the Western world – have faced a new era of mobility and restrictions that they could not have imagined before spring 2020. Although we were caught by surprise by this black swan, we must look forward to short-term recovery and, in the long term, evaluate the consequences and prepare for them. Having said that, Hall *et al.* (2020: 584) see that COVID-19 might have an impact on people's travel behaviour, but add that transforming tourism as a system is difficult (Hall *et al.*, 2020: 584). Nevertheless, we anticipate that social tourism, social entrepreneurship and other social themes may be of great importance in the post-crisis period.

While the COVID-19 crisis clings to all plans and images of the future at the moment, at a more general level, the pre-crisis issues and development goals are still there and are still relevant – indeed, even more so. Scholars have called for more sustainable forms of tourism in the post COVID-19 era (e.g. Haywood, 2020; Higgins-Desbiolles, 2020). Higgins-Desbiolles (2020: 620) states that it is also scholars' task to 'offer contributions to imagine ways tourism can be developed to enable human thriving and ecological recovery'. Haywood (2020: 607) writes that if there are positive sides to this kind of crisis 'it's the on-going, intense, scrutiny of the past, and exploration of more desirable futures'. García-Rosell *et al.* (2020) state, in turn, that due to COVID-19, environmentally friendly concepts as local and virtual tourism will play an important role in the future development of tourism and may offer better opportunities for those who have been excluded from participating in tourism for health, age and economic reasons.

The future as a chronological dimension is quite different from the past and the present. The past has already happened and can thus be described or explained with some accuracy (De Jouvenel, 2018). The present can be seen, felt and experienced at the moment. The future has one positive aspect: we can have an impact on it. As futures researchers often say, we cannot predict the future, but we can make it ourselves. Thus, as a chronological dimension, the future is full of possibilities. It is at the same time known and unknown. As De Jouvenel (2018: 11) writes, we should see

> … the future not as territory to explore but rather as a territory for action, a field of power. Power here serves as a synonym for opportunity, even if all actors do face restrictions, some tied to the plurality of actors, the converging and diverging objectives pursued, the fields in which they operate, and their respective means.

The Structure of the Book

The futures perspective in this book comprises *actors* (Chapters 1 and 2) *methods* (Chapters 3 and 4) and *practices* (Chapters 5 and 6). Action is a significant intervention, because the future has yet to be made. To understand the future, we need different angles – and actors can provide them. In the following chapters, inclusion is examined from the viewpoints of various actors and at many levels of decision-making. Actors that become visible in this book include customers, local residents, entrepreneurs, developers, practitioners and tourism-related industries. Inclusion is built around these actors and entities, which allows profound questions to emerge: can inclusion be produced or organised, or does it require active participation? Inclusion is also knowledge of what exists and how it can be reached. We see knowledge as an integral part of the development of inclusive tourism because we make decisions based on our knowledge of the present. Thus, it is justifiable to claim that tourism inclusion cannot exist without knowledge of the possibilities. We would therefore like to provoke a discussion about the importance of methods in acquiring such knowledge. Furthermore, it is important to draw attention to the practices of actors using their knowledge. This division into actors, methods and practices appears to be an effective way to organise the material, but it is rather theoretical as, in practice, most of the chapters cover all three. The chapters in this book produce multidimensional and timely perspectives on tourism inclusion. These are examples of multiple viewpoints that provide a basis for identifying a holistic approach to inclusive tourism and its futures.

In Chapter 1, Emily Höckert, Outi Kugapi and Monika Lüthje explore the perspective of local residents. They state that participatory development in today's project world is nowhere more important than in inclusive tourism. They note, however, that inclusion also incorporates freedom of choice: that choice can be not to participate but to stay excluded. According to Höckert *et al.*, inclusive development cannot be pre-organised. A certain openness to different ways of being, knowing and doing must remain part of the process. They suggest that development projects should nurture the idea that hosts' and guests' roles are reciprocal and that they change depending on the situation. The authors discuss their own project, Culturally Sensitive Tourism in the Arctic (ARCTISEN), in the context of the thoughts of hosts and guests. This project, financed by the Northern Periphery and Arctic Programme (NPA), aims to help small tourist enterprises in the Arctic develop culturally sensitive products. By reflecting on their own personal experiences, the authors make visible many participatory tourism practices and challenge us to think about who in the project world is the host and who is the guest. They question the restrictions and structures of project-based development and ask whether, in the future, we should have restrictions at all.

New forms based on business models enabled by the sharing economy can play a part in supporting inclusive tourism, but they can also challenge the development of responsible tourism – at least at the destination level. In Chapter 2, Andreja Trdina, Salla Jutila and Maja Turnšek introduce critical thoughts about the inclusiveness of the sharing economy in tourism from a futures perspective. A disruptive change in the operating environment challenges actors in the industry – as well as local residents – by enabling new business models and inviting new target groups and service providers to the table. The authors raise the important issue of emotional and aesthetic capital in relation to the services of the sharing economy. After analysing conversations on the Airbnb online community and its accommodation and service listings, they conclude that being able to benefit from sharing economy platforms requires skills, capital and a certain level of professionalism. Interestingly, sharing economy platforms – even if they widen the scale and efficiency of services and enable the entry of new actors – do not necessarily have a levelling effect. While there is some equality of access, the platforms raise new questions of socially constructed divisions between those who have the skills and means to benefit from them and those who do not.

In Chapter 3, Anu Harju-Myllyaho and Salla Jutila describe their research into tourism strategies. They use a futures research method called causal layered analysis (CLA) to analyse the national and regional tourism strategies of three European countries located at the edge of the continent and well-known as tourist destinations. CLA produces information and knowledge on many layers. First, the authors look at the data, trends and empirical evidence, and describe the countries' strategies at a superficial level. They follow this with a systemic investigation, a global perspective and a consideration of the subconscious layer. Different aspects of inclusion are examined at each stage. As the various layers reveal different elements of the strategies, one begins to understand them more. At the same time, it is possible to outline alternative futures and, more specifically, how inclusion in tourism could be constructed in a different way.

For their framework, Myllyaho and Jutila introduce concepts that are either parallel to or part of the term inclusion. Although inclusion is found in one way or another in each of the strategies, they can be seen as 'children' of their time. Liberal economic thinking, as well as the ethos of growth, is clearly visible. Inclusion is seen as something that it is possible for participants to prepare for, as they need only to enter into the strategy. This is a replication of hospitality – a familiar tourist concept – but at the same time it creates the feeling of a staged situation/participation. This viewpoint is in line with that of Höckert *et al.* (Chapter 1) who describe how, despite decent participatory effort, local people are often put in the position of being guests in projects led by researchers and developers. In more recent tourist strategies, the emphasis has begun to move towards responsible and sustainable aims. For their part, strategies create the future and the way tourism is developed in its destinations, which is why it is important to think about how the future is written.

In Chapter 4, Theres Winter, Seonyoung Kim and Nicola Palmer critically ponder some of the support that has been provided for those in poverty. Their aim is to evaluate the functionality of value chain analysis (VCA) in measuring the impact of pro-poor tourism. The authors conducted their study in the Brazilian municipality of Mata de São João, in the state of Bahia, where tourism has been developed to increase employment and improve people's livelihoods. However, despite good intentions, the poverty rate in the area has stayed relatively high compared with the rest of the country. In their research, Winter *et al.* discovered that VCA is a valid method for analysing tourism income and the involvement of people in poverty in the value chain. A limitation of VCA is that it fails to consider poverty as a holistic phenomenon and thus might limit the method's value as a tool for more inclusive tourism development. Qualitative research to support VCA should, according to Winter *et al.*, focus for instance on clarifying how the political, social, cultural and environmental impacts of tourism influence poverty. This book offers a range of examples of the various complementary methods that can be used and perspectives that can be taken to clarify the impact of tourism and analyse the extent of inclusive tourism.

The importance of inclusive communication is raised by Asunción Fernández Villarán, Mónica Erice, Nagore Espinosa, Ana Goytia, Aurora Madariaga and Ainara Rodríguez in Chapter 5. It is essential that consumers are able to find reliable information easily and to learn whether places and activities are accessible, adapted, inclusive and thus enjoyable for all. The authors explored the information provided concerning accessibility on tourism destination websites. Their analysis covered 147 tourism apps and websites for official tourism bodies in different European countries. They conclude that most of the websites are not particularly user friendly for site visitors with impairments. Much of the information provided is contradictory, out-of-date, inaccurate or lacking.

Inclusion is a multidimensional phenomenon, as noted by Sari Nisula, Marlene Kohllechner-Autto and Krista Skantz in Chapter 6. They argue that social enterprises could enhance social inclusion in sparsely populated areas such as Finnish Lapland. According to the authors, social inclusion is a particular challenge in such places, partly because of residents' isolation from large population centres and their limited access to healthcare or employment. These are important issues, and the authors interviewed people from three of the region's enterprises, asking them about the having, acting and belonging (Raivio & Karjalainen, 2013) of social inclusion in relation to tourism employment, for example. On the matter of social entrepreneurship and social inclusion, the authors conclude that the tourism industry is now in a phase of rapid growth, so opportunities might be made available for those in a difficult labour market position. This would advance social inclusion and provide a much-needed workforce for tourism businesses.

Despite the inherent diversity of the chapters in this book, they share a joint interest in supporting the view that the role of social sustainability – and social inclusion in particular – will become more prominent in the tourism industry. Tourists' values are changing and the demand for more socially sustainable products is growing. We, as authors and editors of this book, are both from Finnish Lapland. We acknowledge that the geographical setting of the book is European. However, as Biddulph and Scheyvens (2018: 585) argue, questions asked by an inclusive tourism approach are not geographically restrictive. We thus appreciate that there is a viewpoint from outside Europe in the book, in Chapter 4 on Brazil, which adds to the European setting and brings – besides the rigorous content – a reminder that inclusion has many faces that might be different in different geographical areas and might be prone to change.

As the chapters of this book emphasise, one of the key elements of social inclusion in the industry is hospitality. Making guests welcome is a central tenet of tourism, adding value and enhancing the tourist experience. Hospitality differentiates the industry from other service industries. However, the structures that we build to greet and serve our guests might also stand in the way of the process of inclusive co-creation. Hospitality is an effective means of building relationships, but it may at the same time create a sense of 'us' and 'them'. The compilation of the chapters in this book indicates that inclusion means the welcoming of different ways of being, doing and knowing, as well as distributing power to all participants in planning and development. Inclusion involves both empowerment and the division of power – and both are stories full of future possibilities.

References

Battour, M., Hakimian, F., Ismail, M. and Boğan, E. (2018) The perception of non-Muslim tourists towards halal tourism: Evidence from Turkey and Malaysia. *Journal of Islamic Marketing* 9 (4), 823–839.

Baum, T. and Hai, N.T.T. (2020) Hospitality, tourism, human rights and the impact of COVID-19. *International Journal of Contemporary Hospitality Management* 32 (7), 2397–2407. https://doi.org/10.1108/IJCHM-03-2020-0242

Biddulph, R. and Scheyvens, R. (2018) Introducing inclusive tourism. *Tourism Geographies* 20 (4), 583–588.

Buhalis, D. and Darcy, S. (eds) (2011) *Accessible Tourism: Concepts and Issues*. Bristol: Channel View Publications.

Butler, R. and Hinch, T. (eds) (2007) *Tourism and Indigenous Peoples: Issues and Implications*. Oxford: Elsevier.

Cambridge Dictionary (2020) Inclusion. See https://dictionary.cambridge.org/dictionary/english/inclusion (accessed August 2020).

Canosa, A. Moule, B.D. and Wray, M. (2016) Can anybody hear me? A critical analysis of young residents' voices in tourism studies. *Tourism Analysis* 21, 325–337.

Darcy, S., McKercher, B. and Schweinsberg, S. (2020) From tourism and disability to accessible tourism: A perspective article. *Tourism Review* 75 (1), 140–144.

De Jouvenel, H. (2018) Futuribles: Origins, philosophy, and practices – anticipation for action. *World Futures Review* 11 (1), 8–18.

García-Rosell, J.C., Haanpää, M., Hakkarainen, M. and Saraniemi, S. (2020) Covid-19 as a trigger for more responsible tourism business models. See https://businessandsociety.org/covid-19-as-a-trigger/ (accessed August 2020)

Hall, M., Scott, D. and Gössling, S. (2020) Pandemics, transformations and tourism: Be careful what you wish for. *Tourism Geographies* 22 (3), 577–598, doi:10.1080/14616 688.2020.1759131.

Haanpää, M., Hakkarainen, M. and Harju-Myllyaho, A. (2018) Katsaus yhteiskunnalliseen yrittäjyyteen matkailussa: Osallisuuden mahdollisuudet pohjoisen urbaaneissa paikallisyhteisöissä (Review of social entrepreneurship in tourism: Possibilities of participation in the Northern urban communities). *The Finnish Journal of Tourism Research* 14 (2), 44–58.

Harju-Myllyaho, A. (2018) Towards accessible hospitality: Intersectional approach to rainbow tourism. Master's thesis. Rovaniemi: University of Lapland. http://urn.fi/URN:NBN:fi:ula-201802131036

Haywood, M. (2020) A post COVID-19 future – Tourism reimagined and re-enabled. *Tourism Geographies* 22 (3), 559–609, doi:10.1080/14616688.2020.1762120.

Heinonen, S. (2013) Mustien joutsenten tanssi. In *Mustat Joutsenet. Mikä Muuttaa Maailmaa Seuraavaksi?* Tulevaisuusvaliokunnan julkaisuja. 4/2013 (pp. 15–34). Helsinki: Tammi.

Higgins-Desbiolles, F. (2020) Socialising tourism for social and ecological justice after COVID-19. *Tourism Geographies* 22 (3), 610–623, doi:10.1080./14616688.2020. 1757748.

Höckert, E. (2015) Ethics of hospitality: Participatory tourism encounters in the northern highlands of Nicaragua. PhD dissertation, University of Lapland.

Inayatullah, S. and Black, P. (2020) Neither a black swan nor a zombie apocalypse: The futures of a world with the Covid-19 coronavirus. *Journal of Futures Studies. Perspectives.* See https://jfsdigital.org/2020/03/18/neither-a-black-swan-nor-a-zombie-apocalypse-the-futures-of-a-world-with-the-covid-19-coronavirus/ (accessed April 2021).

Isola, A.-M., Kaartinen, H., Leeman, L., Lääperi, R., Schneider, T., Valtari, S. and Keto-Tokoi, A. (2017) *Mitä osallisuus on? Osallisuuden viitekehystä rakentamassa.* Working paper. Helsinki: Finnish Institute for Health and Welfare.

Jafari, J. and Scott, N. (2014) Muslim world and its tourisms. *Annals of Tourism Research* 44, 1–19.

Klemm, M.S. (2002) Tourism and ethnic minorities in Bradford: The invisible segment. *Journal of Travel Research* 41 (1), 85–91.

McCabe, S. (2009) Who needs a holiday? Evaluating social tourism. *Annals of Tourism Research* 36 (4), 667–688.

McCabe, S. and Johnson, S. (2013) The happiness factor in tourism: Subjective well-being and social tourism. *Annals of Tourism Research* 41, 42–65.

Minneart, L., Maitland, R. and Miller, G. (2009) Tourism and social policy: The value of social tourism. *Annals of Tourism Research* 36 (2), 316–334.

Nielsen, K. (2013) Approaches to seniors' tourist behavior. *Tourism Review* 69, 111–121.

Nielsen, N. and Wilson, E. (2012) From invisible to indigenous-driven: A critical typology of research in indigenous tourism. *Journal of Hospitality and Tourism Management* 19, 26–34.

Pramod, K.M. (2020) COVID-19, Black Swan events and the future of disaster risk management in India. *Progress in Disaster Science* 8.

Raivio, H. and Karjalainen, J. (2013) Osallisuus ei ole keino tai väline, palvelut ovat! Osallisuuden rakentuminen 2010-luvun tavoite- ja toimintaohjelmissa. In T. Era (ed.) *Osallisuus – oikeutta vai pakkoa?* (pp. 12–34) Jyväskylän ammattikorkeakoulun julkaisuja 156. Jyväskylä: Jyväskylä University of Applied Sciences.

Rubin, A. (2004) Tulevaisuudentutkimus tiedonalana. TOPI – Tulevaisuudentutkimuksen oppimateriaalit. Tulevaisuuden tutkimuskeskus, Turun yliopisto. See https://tulevaisuus.fi/perusteet/tulevaisuudentutkimus-tiedonalana/ (accessed March 2020).

Simmons, D. (1994) Community participation in tourism planning. *Tourism Management* 15 (2), 98–108, doi:10.1016/0261-5177(94)90003-5

UNESDOC (1982) International Year of Disabled Persons (IYDP). See https://unesdoc.unesco.org/ark:/48223/pf0000048396 (accessed August 2020).

Yang, L. and Wall, G. (2009) Ethnic tourism: A framework and an application. *Tourism Management* 30 (4), 559–570.

Zeppel, H. (2007) Indigenous tourism. In C. Game (ed.) *Tourism Planning and Policy* (pp. 403–411). Wiley.

Zsarnoczky, M., David, L., Mukayev, Z. and Baiburiev, R. (2016) Silver tourism in the European Union. *GeoJournal of Tourism and Geosites* 18 (2), 224–232.

Part 1
Actors

1 Hosts and Guests in Participatory Development

Emily Höckert, Outi Kugapi and Monika Lüthje

Introduction

During the past decades, the idea of local participation has played an important role in the search for more sustainable, responsible and inclusive ways of developing tourism. The basic idea behind the participatory approach is to guarantee local communities' active involvement in their own development. In practice, the initiatives for inviting more tourists and enhancing tourism development quite often come from outsiders. Various examples indicate that despite, and even because of, the good intentions of enhancing inclusion and well-being, local communities tend to play the role of the guests in participatory projects hosted by researchers and development practitioners (see Butcher, 2007).

The participatory approach can be located at the core of 'inclusive tourism' as it aims to ensure that marginalised groups can take part in consuming, producing and sharing the benefits of tourism activities (Scheyvens & Biddulph, 2017). In Scheyvens and Biddulph's (2017) view, the idea of inclusion consists of two basic aspects – first, who are included or excluded in tourism and, second, on what terms. These questions have been discussed in the context of inclusive business growth (Hall *et al.*, 2012), accessible tourism (Buhalis & Darcy, 2011; Darcy, 2010), social tourism (Minnaert *et al.*, 2011), labelling processes (de Bernardi *et al.*, 2018), social justice (Aitchison, 2007; Jamal, 2019), social entrepreneurship (Haanpää *et al.*, 2018) and digitalism (Minghetti & Buhalis, 2010). While searching for new ways to enhance inclusive tourism practices, these studies underline the importance of participating in tourism activities based on one's own conditions, needs and interests (de Bernardi *et al.*, 2018; George *et al.*, 2009; Jamal & Dredge, 2014: 195–197; Müller & Viken, 2017; Schilcher, 2007: 59). In other words, the idea of inclusive and participatory tourism development should also contain the possibility of free choice to not participate; that is, to remain 'excluded' from tourism projects.

As a concept, *project* emphasises agency, plan, objectives, volition and accomplishment (Rantala & Sulkunen, 2006a: 8–9). Participation in projects is organised through collaboration, partnerships and agreements that structure the relationships of various project participants and are based on voluntary and mutual commitment, negotiations and trust (Sulkunen, 2006: 17–18; see also Lundin, 2016; Ren et al., 2018: 181). However, one of the persistent challenges in today's 'project society' (e.g. Lundin, 2016; Sulkunen, 2006), where strategies and financing for participation often come from external actors, is what happens when the support ends and the projects are handed on to local stakeholders (see Zapata et al., 2011). 'The project' can also be seen as a neoliberal solution where individuals are expected to develop innovative and entrepreneurial solutions to structural problems (Rantala & Sulkunen, 2006b; Sulkunen, 2006; see also Lundin, 2016). Indeed, critical examinations of community-based projects indicate how the principle of participation does not automatically lead to more equal power relations between different actors (Butcher, 2007; Höckert, 2018; Wearing & Wearing, 2014). It seems that, despite the good intentions of enhancing people's ownership in their own well-being, the 'project society' is in constant need of structural changes and tuning in order to secure partners' commitment and ownership within participatory projects.

The purpose of this chapter is to approach inclusion by discussing the roles of hosts and guests in participatory tourism projects. Instead of drawing inspiration from the predominant understanding of host–guest relations within hospitality management (see Lashley, 2017), we call attention to the more 'ancient' idea of hospitality, where – in its simplest form – hosts have the responsibility to take care of their guests' well-being for a limited amount of time (O'Gorman, 2010). Moreover, in the context of 'project society', we are not focused on host–guest relations that take place in different kinds of physical homes, but approach projects as metaphorical homes where different kinds of moments and relations of hospitality occur (see Germann Molz & Gibson, 2007; Höckert, 2018).

Instead of celebrating all the participants as 'the hosts', we draw explicit attention to structural challenges of our project worlds and to the ways in which the host–guest roles keep changing during the project processes. To visualise and demonstrate our approach in practical terms, we wrote this chapter side by side with a development project called Culturally Sensitive Tourism in the Arctic (ARCTISEN). Our aim was to weave together the literature on participatory development and hospitality with our own experiences in preparing this project. In addition to our reflective memory work, the analysis draws on a wide range of documentation from the preparatory phase, such as meeting memos, email correspondence, and reports and documents from the funding authority.

Preparation of the ARCTISEN project was driven by our interest to enable small and medium-size tourism enterprises to visit and learn from each other and to co-create culturally sensitive tourism products (ARCTISEN, 2018). The very first step of our project journey was taken in 2015, when Monika Lüthje proposed the idea of an Indigenous tourism project to the rest of us. From the very first stages, she opened the door for shared hostessing (see also Veijola & Jokinen, 2008) of all the new ideas that began to arrive. We decided to apply for funding from the EU's Northern Periphery and Arctic Programme (NPA, 2018). During the preparatory phase of the ARCTISEN project, our role was to learn and follow the conditions set by our NPA 'host'. We acknowledge that our affiliation with the University of Lapland made us look like mature guests with a well-established reputation for being able to 'follow the rules' (Germann Molz, 2014; Lundin, 2016). Nevertheless, while being the guest knocking on NPA's doors, our university team was also taking on the role of the host, who began to welcome tourism entrepreneurs, destination management organisations (DMOs), non-governmental organisations (NGOs), municipalities and other university partners to join the preparation phase of the project. By doing this, we wished to form a new 'tourism knowledge collective' (Ren & Jóhannesson 2018: 24): a gathering around culturally sensitive tourism.

The next section guides the reader along the streams of discussions on participatory development within *tourism studies*. From there, we move to our theoretical take on host–guest relations. The next section describes how we then laid our hopes on the NPA as the host for our transnational project idea. In the final section, we conclude this chapter and suggest that the idea of hosts and guests can be used as a fruitful approach when envisioning and promoting alternative, more inclusive, sensitive and responsible tourism futures.

Participatory Development in Tourism

While the history of 'participation' – of being, doing and knowing together – is as old as humanity, it has become both a keyword and a buzzword in the contemporary search for sustainable development (Berkhöfer & Berkhöfer, 2007; Cornwall, 2006; Stiefel & Wolfe, 1994). Originally, the emphasis on active local participation emerged as a response to the numerous tourism impact studies and resident attitude surveys, which indicated that few positive impacts accrued to host communities (Cohen, 1979; Keogh, 1990: 450; Tosun, 2000: 616). In the late 1970s and early 1980s, the tourism sector was still marked by little public involvement in tourism planning and it was noticed that public concerns should be incorporated into decision-making processes (de Kadt, 1979; Mathieson & Wall, 1982). Ever since, the idea of local participation has been connected, most of all, to small-scale tourism development that uses cultural and environmental resources in responsible and sustainable ways (Jamal & Getz, 1995; Scheyvens, 2002; Tuulentie & Sarkki, 2009).

Researchers have since been formulating alternative development approaches, such as *community-based tourism*, which focuses on the well-being of local host communities (see Höckert, 2011; Höckert *et al.*, 2013; Jamal & Dredge, 2014; Saarinen, 2006, 2010; Telfer, 2009; Tuulentie & Sarkki, 2009). The term *Indigenous tourism* has also been seen as a form of tourism that actively involves Indigenous communities in activities and decision-making and/or acting as an attraction of the area (Hinch & Butler, 1997: 9; Hinch & Butler, 2009: Müller & Viken, 2017; see also Kugapi & de Bernardi, 2017). It has been argued that, for many Indigenous people, tourism is an opportunity to earn extra income, show others part of their culture, disseminate knowledge (Tuulentie, 2006) and gain community control and ownership of tourism. Good examples of previous projects in tourism that have been planned and led by Indigenous and other local communities include the cultural and environmental *Sápmi Experience* that was created by VisitSápmi in Sweden (see de Bernardi *et al.*, 2018) and the guidelines for responsible and ethically sustainable Sámi Tourism produced by the Sámi Parliament (2018) in Finland.

In recent decades, a growing number of governments and international development agencies have come to recognise the important role of local-level organisations and local-level knowledge. As a result, the participatory discourse has played an important role in the language used in project plans. Simultaneously, Jim Butcher (2007) and other tourism scholars have presented sharp critiques towards participatory tourism projects run by NGOs and aid agencies. Butcher (2007) argues that while community participation is often associated with a progressive democratic approach, communities are invited to participate only to implement pre-planned projects rather than in shaping the development goals and agendas behind them. This is especially true for Indigenous communities who are often seen as 'targets rather than as agents of development' in tourism projects (Müller & Viken, 2017: 7). Moreover, critical voices within tourism studies have also drawn attention to the negative influences of participatory tourism development, such as problems in achieving the goal of benefit delivery, aggravating and creating internal conflicts and jealousies, and promoting unrealistic expectations (Hinch & Butler, 1996; Müller & Viken, 2017; Swarbrooke, 2002; Tosun, 2000; Warnholz & Barkin, 2018).

The failure of supposedly participatory projects has been explained by top-down approaches that overlook local contexts and local knowledge. In these kinds of projects, external actors – the guests – arrive in communities with ready-made plans and ideas on how the local actors should participate in their own development (see Höckert, 2018; Jamal & Dredge, 2014). As an alternative to top-down methods, bottom-up strategies place emphasis on ownership and empowerment that can lead to social, economic, psychological and political change (Arai, 1996; Scheyvens, 1999; 2002, 2003; Telfer & Sharpley, 2008: 130). Wearing and Wearing (2014) approach the issue of moral encounters in ecotourism from a feminist post-colonial perspective, which directs attention to the inequalities and intersections of gender, race and socioeconomic positions within host communities. The conceptualisation of empowerment has been used in research that focuses on the issues of gender equality (Hashimoto, 2014: 223–225; Miettinen, 2007) and Indigenous issues (de Bernardi *et al.*, 2018; Nicholas & Thapa, 2018), in the context of tourism development. It has been argued that external contacts, self-esteem, pride and confidence can have a positive influence on empowerment, whereas a lack of knowledge about tourism, a lack of self-confidence or a lack of skills might lead to disempowerment even though people are seemingly participating in tourism development (de Bernardi *et al.*, 2018; Höckert, 2011).

In recent years, an increasing amount of post-development literature has questioned the dominance of Eurocentric worldviews on development and called for the inclusion of multiple worldviews and ways of understanding tourism and development in general (Telfer, 2009). Therefore, the notions of *local* and *Indigenous knowledge* have become commonly used concepts within the participatory tourism discourse (Jamal *et al.*, 2003: 154; Prasetyo *et al.*, 2019: 14; Telfer, 2009: 153; Zapata *et al.*, 2011: 23). What much of the previous studies seem to agree on is how local stakeholders hold essential knowledge (Lee & Jan, 2019; Lundberg, 2015; Tanga & Maliehe, 2011; see also Kaján, 2014) that should be included in tourism development from the early stages (Lee & Jan, 2019). Nevertheless, while some researchers call for more careful attention to local knowledge (Koster *et al.*, 2012), others argue that local communities are often lacking the needed knowledge and are thus seriously hindered from participating in planning and developing tourism (see Moscardo, 2008; Warnholz & Barkin, 2018). For instance, Tosun (2000: 630) has suggested that difficulties can be explained by '...cultural remoteness of host communities to tourism-related businesses in developing countries...' or local communities' unawareness of tourism markets. At the same time, Hakkarainen (2017) and Höckert (2018) have drawn attention to local actors' limited time and other resources to participate in project activities outside of their usual daily routines. In their view, this challenge has been overlooked in both research and practice.

Despite the critique and scepticism, development scholars encourage others to be careful not to throw the baby out with the bathwater and discard the idea of participation as such (e.g. Hickey & Mohan, 2004; Leal, 2010: 77). For instance, in participatory scholar Leal's view, there exists a need to return to alternative constructs of 'the good life' (Leal, 2010: 79). In an extensive critique of the participatory 'orthodox' in tourism studies, Butcher (2007: 61) laments that even the comprehensive critical studies tend to focus on operationalising the concept of community participation rather than on the concept itself. Butcher (2012: 103) argues, similar to Wearing and Wearing (2014), that studies on local participation are often misleadingly focused only on inclusion inside the local communities and not the power relations beyond the community level. Despite the many participatory development and research projects, there has been little reflection about the value premises that shape our opinions of ideal forms of participation and development as such. For us, one of the consequences of the extensive focus on methodological packages and techniques is that the conceptualisation of community participation has lost its connections to previous theories of community development and participation, and participation has lost its philosophical meanings (Leal, 2010; see also Dredge *et al.*, 2013; Jamal & Stronza, 2008).

On Hosts and Guests

Previous and existing challenges in participatory projects encourage looking for novel ways of thinking about tourism development to take into consideration a wider range of stakeholders and to sensitise ourselves towards Indigenous and other local cultures. During recent years, we have become convinced about the fruitfulness of approaching participatory development in terms of host–guest relations. This means, in its simplest form, replacing the goal-driven and growth-driven ideas of participation with a call for openness and reciprocity between hosts and guests (Germann Molz & Gibson, 2007; Höckert, 2018; Keen & Tucker, 2012: 97). The idea of reciprocity between hosts and guests can be understood not only as a ritual of exchanging gifts, but also as a more fundamental care relationship, where both hosts and guests take care for each other's well-being (see Länsman, 2004; Lashley, 2000; Pyyhtinen, 2014; Telfer, 2000).

In ancient stories, hospitality was described as the virtue of opening one's home to a stranger who arrives at the door (O'Gorman, 2010). This refers to the responsibility of welcoming and taking care of the one in need. Hence, the idea of hospitality simultaneously includes the call for openness towards strangers and the responsibility to offer them what they might need. However, the responsibility to take care of one's guests came with no guarantee that the surprise guest would be able to 'pay back' the hospitality of the host. The only thing that could be expected from this guest, as Immanuel Kant (1996 [1795]) later described, was not to take advantage of or abuse the host's hospitality.

The notion of hospitality has gained attention in recent years based on the growing mobility of migrants, asylum seekers, tourists, commodities and so on (Lynch *et al.*, 2011). Today's tourism industries have turned hospitality into a profitable business, where the idea of reciprocity means that 'guests' pay for the hospitality services that their 'hosts' offer (see Smith, 1977). While this aspect of hospitality has taken over a big part of tourism research and education, our theoretical idea of hospitality builds on the philosophies of hospitality where the focus is on the questions of ethics, responsibility and care among hosts and guests (Germann Molz & Gibson, 2007; Lynch *et al.*, 2011; Veijola *et al.*, 2014). These streams of discussions share the idea of keeping the roles of hosts and guests on the move (see Derrida, 1999; Levinas, 1969), reflecting upon the ways in which we are constantly both hosts and guests in our relations with others. What makes this approach especially fruitful is the way in which this kind of hospitality – read participation – can never be completely regulated or pre-planned; instead, it is continuously negotiated in the encounters between self and other (Höckert, 2018).

In a recent analysis, Tucker (2014: 199) points to the need for moving away from the 'assumptions of fixed cultural positions in tourism encounters, and towards focusing on the fluidity and mobility of positions and relation between so-called "tourists" and "toured"', hosts and guests. In our view, this paradigm shift is essential as it places the focus on the contradictions and ambiguities of different tourism encounters. Saying this, we suggest here that the notion of hospitality and the idea of caring relationships between hosts and guests can help us to reflect different ways of thinking, doing and accomplishing participatory development (see Höckert, 2018). As proposed by Levinas (1969), hospitality boils down to the idea of being ready for surprises and keeping the door open to the unexpected. Along these lines, we see that participatory development cannot be pre-designed, but must remain open to other ways of being, doing and knowing. Moreover, and still following Levinas' radical thought, instead of trying to preserve our roles as hosts of participatory projects, we must be ready to let others take on the role of the hosts; that is, being a guest is supposed to be a temporary position as it would be unbearable to always be in the role of a guest needing to follow the conditions and 'house rules' of one's host. Therefore, we suggest here that participatory projects should strive for reciprocal relations, where the roles of hosts and guests are constantly changing.

This is something that we wish to demonstrate in the following section and to introduce an alternative way of approaching subjectivity and agency in participatory projects.

Hosts and Guests in the ARCTISEN Project

The idea for the ARCTISEN project grew out of concerns about the exploitation of Sámi and other Indigenous cultures in the middle of an expansive growth in tourism beginning in 2015. While acknowledging previous Sámi tourism development projects in the area, the need for a more comprehensive, international project was supported by previous research on Indigenous Sámi tourism in northern Norway, Sweden and Finland and the Kola Peninsula in Russia (e.g. de Bernardi *et al.*, 2018; Lüthje, 1998; Müller & Huuva, 2009; Müller & Pettersson, 2006; Niskala & Ridanpää, 2016; Viken & Müller, 2017). The starting point for ARCTISEN was to compare current situations across the borders, learn from others and develop something new, while the project participants, at the same time, become more powerful against the exploiting tourism industry. With a large, participatory project in mind, we began the preparatory phase to apply for project funding from the NPA.

The NPA is an EU development programme with the vision to 'generate vibrant, competitive and sustainable communities, by harnessing innovation, expanding the capacity for entrepreneurship and seizing the unique growth initiatives and opportunities of the Northern and Arctic regions in a resource efficient way' (NPA, 2018: 2). The NPA makes open, public calls for project proposals to act as hosts who invite guests (read project applicants) to suggest what kind of development projects should take place in the programme area (Finland, Sweden, Norway, Iceland, Greenland, Faroe Islands, Ireland, Northern Ireland and the United Kingdom) within the frame of its development programme. The NPA offers the possibility to apply for funding to prepare the actual (main) project application. We received preparatory funding in May 2016, which allowed us to welcome and include more stakeholders to co-plan the project.

The initial idea was to have a Sámi tourism project including Finland, Sweden and Norway. However, to fulfil the requirements of our host, the NPA, we had to enlarge our project to a more transnational one. The NPA recommended that we invite more partners to join our project, such as the World Indigenous Tourism Alliance and tourism organisations in Greenland and Canada. We aimed at a project with several kinds of partners – universities, business development agencies, small and medium-size enterprises, DMOs and NGOs. We, at the University of Lapland, as a lead partner, acted as a host and invited the previously mentioned organisations to participate and contribute to the project preparation as our guests. Our main criteria was to find guests who were sharing our concerns about the cultural insensitivity of current tourism development in the Arctic. While our focus was on the responsible use of cultures in tourism settings, we chose to make the project more inclusive by speaking of *culturally sensitive tourism* rather than *Indigenous tourism*. We defined culturally sensitive tourism as tourism that enhances stakeholders' self-determination, intra- and intercultural understanding, respect, empowerment and inclusion together with economic development.

What is important to acknowledge here is that the NPA encourages projects to invite two different kinds of partners – full partners and associated partners. Full partners receive funding from the NPA for project implementation; they participate in project funding with their own contribution and are responsible for project activities and reporting. Interestingly, only a community that is a legal entity – an organisation – can be a project partner in NPA projects, not, for example, a community formed by local people without membership in a formal organisation. In addition, while associated partners do not receive funding from the NPA or participate in project funding or reporting, they can participate in project activities and get their project costs covered from the project budgets of the full partners. This made it possible for smaller organisations with limited resources to join ARCTISEN as associated partners. However, while the different kinds of partnerships enable a wider range of actors to join and participate in project activities, this structure excludes the smaller associated partners from project management and decision-making processes, resulting in unequal power relations within the project.

The aim of the ARCTISEN project was to find and develop solutions for the different needs of small and medium-size tourism enterprises in the Arctic. We aligned ourselves with the idea of pioneering tourism researcher Emmanuel de Kadt that 'for community interests to be taken into account in tourism (or any other) development, it is essential that those interests be articulated from the moment potential projects are identified' (de Kadt, 1979: 134). Moreover, we took seriously the aforementioned criticisms of Butcher *et al.* (2012: 118) towards the paradoxes within the participatory tourism paradigm and the misuse of participatory rhetoric. We aimed to be responsible and respectful hosts when we arranged interviews and discussions with local stakeholders, respecting their rhythms, timelines and interests. While we invited these actors to join the project planning, we were simultaneously guests who were entering their premises – thus making them the hosts.

During the discussions, our aim was to learn about the challenges, possibilities and needs of development that the local stakeholders currently faced in tourism (see Kaján, 2014; Lee & Jan, 2019; Lundberg, 2015; Tanga & Maliehe, 2011). We also discussed what kind of project they would like to have and what activities to include. As Hall (2003: 100) suggests, tourism planners have the task of finding agreement between various stakeholders and interests in tourism development. At the same time, we invited the stakeholders to participate in the project proposal as project partners and become co-hosts instead of guests. And this also happened: the organisations we invited as project partners invited new partners to the project – partners they deemed relevant.

Along the NPA's horizontal principle of inclusion and diversity (NPA, 2018: 10), we decided to keep the 'project home' open to everybody interested in culturally sensitive tourism development. We also wanted to plan the project so that the project activities were not only for the project partners, but others interested in them may also participate. Planning the project together with a large number of different kinds of stakeholders was a learning process for all of us (e.g. Grimwood *et al.*, 2012; Koster *et al.*, 2012). In line with the idea of hospitable forms of participation, the project plan became quite different from our initial ideas. This was not solely due to our discussions with the stakeholders and input from our project partners, but also because of the requirements of the NPA concerning the objectives, structure, contents and partners of the project.

After an extensive preparatory phase, we received both disappointing and encouraging news: while our first project proposal was not approved, the NPA encouraged us to modify our application and re-apply for funding. According to the NPA, one of the many challenges with the proposal was the high number of project partners, which would have made the project difficult to manage. It seemed like we had been too inclusive and welcoming during the process, and were hence urged to cut down the number of project partners. This happened quite organically: while preparing the second draft of the project plan, some of the former project partners decided to drop out due to lack of staff and/or financial resources. NPA projects have to cover part of the project costs themselves, which is normally done by allocating working hours of permanent staff to the project. Preparation of the ARCTISEN project proposal took a lot of time and effort, and required the skills and prior experience of EU projects. In our view, it seems that only large organisations have the capacity to prepare this type of transnational project proposal. It also has to be said that, without the preparatory project funding, our university would not have been able to invest so much labour in the project planning and we would not have been able to involve so many stakeholders in the project preparation as we now could.

Interestingly, the possibility of choosing *not* to participate in tourism development is rarely discussed in the academic debates on local participation in tourism. While the participatory tourism literature takes for granted communities' interest in participating, Schilcher (2007: 59) and Jamal and Dredge (2014: 195–197) (also see Butcher, 2007, 2012: 104; George *et al.*, 2009; Jamal & Stronza, 2009) are among the few authors who have brought up the question of whether people can choose not to participate in tourism development. In other words, this means accepting that tourism is not always perceived as an activity that adds to the general well-being of local communities, or to particular individuals' well-being within those communities. The core of Hinch and Butler's (1997) definition of Indigenous tourism is that Indigenous communities should have the opportunity to choose whether and how they want to be involved in tourism (see Müller & Viken, 2017). In our case, all the stakeholders invited to participate in the project were those who already lived off and with tourism (see Ren & Jóhannesson, 2018: 24). However, we do not know how many of them did not become our project partners because culturally sensitive tourism was not the kind of tourism they wanted to develop or because we (or the NPA) were not the kind of hosts with whom they would have liked to develop it. In any case, all those invited had the freedom to choose whether to participate in the project or not. We respected those decisions and did not ask for explanations.

The second version of our application was approved, which enabled us to start with the actual ARCTISEN project in October 2018. The inclusion principle of the NPA, the different partnership forms and the preparatory funding offered by the programme – as well as keeping the roles of hosts and guests changing – allowed us to prepare a project that includes various kinds of engaged stakeholders as partners. Although we were constrained by various administrative and practical matters stemming from the funding programme and the partner or other stakeholder organisations, we ensured that the project was built on stakeholders' needs.

Swarbrooke (2002: 128) noted that, as community involvement in tourism planning can slow down and add costs to tourism planning, it can lead to faster top-down strategies. However, one commonly identified problem in community-based tourism projects is that the development brokers or tour operators might enter rural areas without prior understanding of the local realities or, for instance, the interconnection between tourism and community development (Wearing & McDonald, 2002). In practice, this has led to the implementation of participatory projects in which local communities are not properly informed about what they are participating in and what impacts their participation may have (Sammels, 2014). We agree. The participatory approach requires resources: time, money, people and cultural sensitiveness, among others.

To avoid these problems, we planned the first phase of the project to be a research phase involving interviewing project stakeholders to further improve our understanding of their challenges, development needs, visions and wishes concerning the project. The rest of the planned project activities were based on this information, many of them co-created together with the stakeholders participating in the project. However, the project activities must be congruent with the plan we presented in the project application. From a participatory perspective, we find this problematic, especially in the rapidly changing tourism business where new challenges and development needs may be unanticipated. This is a constraint we have to negotiate with the NPA during the project implementation phase to keep the project up-to-date and inclusive.

Towards Inclusive Project Futures

The purpose of this chapter was to rethink the idea of enhancing inclusive tourism development through participatory tourism projects. We have suggested here that the notions of hosts and guests can offer an alternative way of understanding the pitfalls and possibilities of initiatives that aim at giving voice to a wide range of tourism stakeholders. Keeping the roles of hosts and guests constantly changing in participatory project preparation and implementation may result in more inclusive development projects. Therefore, we suggest that participation in project development can be thought as taking the roles of hosts and guests who care for each other's well-being through the project process – and even after (see Ren & Jóhannesson 2018).

We have argued that the idea of participation meets many practical constraints and limitations that must be taken into account to enable genuine inclusion, involvement and engagement (e.g. Hall, 2003; Scheyvens, 2011). While entrepreneurs and other tourism stakeholders have only limited time and other resources (Hakkarainen, 2017; Höckert, 2018), participation – as host and/or guest – requires resources and meeting various conditions set by the funding bodies. Nevertheless, we hope to avoid the unfruitful either–or debate on whether or not we should do participatory projects or whether or not local communities should be included in tourism development. In our view, this kind of discourse should be avoided as it keeps constructing an illusion of local participation as something decided and controlled by outsiders – as if participation and inclusion were something that could be initiated or stopped merely by external experts.

Although there is consensus among tourism scholars that tourism and development brokers play a significant role in participatory tourism projects (e.g. Cheong & Miller, 2000; van der Duim *et al.*, 2006), opinions about the responsibilities of these development intermediaries vary greatly. In addition to distributing financial support to participants, project workers should also provide technical assistance, capacity building and possibilities for networking (see Miettinen, 2007; Wearing & McDonald, 2002; Wearing & Wearing, 2014). While planning and preparing for the ARCTISEN project, we aimed at creating a 'project community' in which the roles of teachers and learners, hosts and guests would be fluid and constantly changing. Indeed, project type development can be thought of in terms of formulating temporary communities of hosts and guests. However, instead of trying to simplify or categorize projects or local communities we should see them (us) as inevitably complex and diffuse, and continuously on the move (see Cole, 2006: 95; Veijola *et al.*, 2014).

In this chapter, we have considered projects as things that can never be completely regulated or pre-planned but continuously negotiated in the encounters between self and other. To re-think the relations between tourism experts and different kinds of tourism stakeholders, we suggest questioning goal-driven ideas of participation and call for new spaces for more mobile and hybrid subject positions (Höckert, 2018; Keen & Tucker, 2012: 97; Tucker, 2014). Saying this requires the deconstruction of assumed or pre-defined roles of hosts and guests, teachers and learners in participatory development (for the notion of post-host–guest society, see Veijola & Jokinen, 2008). Moreover, we aimed to challenge the idea that project workers alone have the role of the hosts, responsible for planning and arranging the 'best party' ever.

While imagining more hospitable and innovative ways of doing development, we have drawn attention to the challenges of hospitable projects within the prevailing project funding systems – with precise plans and measurable outputs. As shown by the example of our preparation for the ARCTISEN project, funding schemes can both enable and constrain inclusion in development projects. In our case, the preparatory project funding as well as the possibility to include associate partners in the project and offer project activities to others than the project partners made the project more participatory and inclusive. At the same time, only formal organisations were accepted as project partners and, in order to receive funding, they had to contribute their own resources to the project as well. Are these constraints necessary? Could we imagine an alternative kind of project society?

While writing the last lines of this chapter, the ARCTISEN project has already been running for a year. At this point, we are co-hosting the project with other partners who seem to have strong engagement and ownership in the project. The hosting and guesting of the project activities has been shared among the project partners and a wide range of other tourism stakeholders in a way that could not be completely pre-planned or anticipated. Instead of expecting that all the doors would be open or opened for us, we are committed to continue culturally sensitive negotiations with a wide range of hosts and guests.

Acknowledgements

Many thanks to all the small and medium-size enterprises, municipalities, DMOs, academics and others who believed in the ARCTISEN project idea and participated in developing it further. We also want to express our gratitude to all our colleagues at the University of Lapland and other institutes and organisations who contributed to construction of the project's foundations over the years. Not least of which, we wish to thank the people within the NPA for being supportive in their hosting and the referees who commented on our chapter. Thank you!

References

Aitchison, C.C. (2007) Marking difference or making a difference: Constructing places, policies and knowledge of inclusion, exclusion and social justice in leisure, sport and tourism. In I. Ateljevic, A. Pritchard and N. Morgan (eds) *The Critical Turn in Tourism Studies: Innovative Research Methodologies* (pp. 77–90). Amsterdam: Elsevier.

Arai, S. (1996) Benefits of citizen participation in a healthy communities initiative: Linking community development and empowerment. *Journal of Applied Recreation Research* 21, 25–44.

ARCTISEN (2018) See http://sensitivetourism.interreg-npa.eu/ (accessed March 2020).

Berkhöfer, U. and Berkhöfer, A. (2007) 'Participation' in development thinking – Coming to grips with a truism and its critiques. In S. Stoll-Kleemann and M. Welp (eds) *Stakeholder Dialogues in Natural Resource Management: Theory and Practice* (pp. 79–116). London: Springer.

Buhalis, D. and Darcy, S. (eds) (2011) *Accessible Tourism, Concepts and Issues*. Bristol: Channel View Publications.

Butcher, J. (2007) *Ecotourism, NGOs and Development*. London: Routledge.

Butcher, J. (2012) The mantra of 'community participation' in context. In T.V. Singh (ed.) *Critical Debates in Tourism* (pp. 102–108). Bristol: Channel View Publications.

Butcher, J., Weaver, D. and Singh, S. (2012) Does community participation empower local people? In T.V. Singh (ed.) *Critical Debates in Tourism* (pp. 101–121). Bristol: Channel View Publications.

Cheong, S. and Miller, M. (2000) Power and tourism: A Focauldian observation. *Annals of Tourism Research* 27 (2), 371–390.

Cohen, E. (1979) A phenomenology of tourist experiences. *Sociology* 13 (2), 179–201.

Cole, S. (2006) Cultural tourism, community participation and empowerment. In M.K. Smith and M. Robinson (eds) *Cultural Tourism in a Changing World: Politics, Participation and (Re)presentation* (pp. 89–103). Clevedon: Channel View Publications.

Cornwall, A. (2006) Historical perspectives on participation in development. *Commonwealth and Comparative Politics* 44 (1), 27–43.

Darcy, S. (2010) Inherent complexity: Disability, accessible tourism and accommodation information preferences. *Tourism Management* 31 (6), 816–826.

de Bernardi, C., Kugapi, O. and Lüthje, M. (2018) Sámi indigenous tourism empowerment in the Nordic countries through labelling systems. Strengthening ethnic enterprises and activities. In I.B. de Lima and V. King (eds) *Tourism and Ethnodevelopment. Inclusion, Empowerment and Self-determination* (pp. 200–212). Abingdon: Routledge.

de Kadt, E. (1979) *Tourism, Passport to Development?* Oxford: Oxford University Press.

Derrida, J. (1999) *Adieu to Emmanuel Levinas*, translated by P.A. Brault and M. Naas. Redwood City, CA: Stanford University Press.

Dredge, D., Hales, R. and Jamal, T. (2013) Community case study research: Researcher operacy, embeddedness and making research matter. *Tourism Analysis* 18 (1), 29–43.

George, E.W, Mair, H. and Reid, D.G. (2009) *Rural Tourism Development: Localism and Cultural Change*. Bristol: Channel View Publications.

Germann Molz, J. (2014) Camping in clearing. In S. Veijola, J. Germann Molz, O. Pyyhtinen, E. Höckert and A. Grit (eds) *Disruptive Tourism and its Untidy Guests. Alternative Ontologies for Future Hospitalities* (pp. 19–41). New York: Palgrave MacMillan.

Germann Molz, J. and Gibson, S. (2007) Introduction: Mobilizing and mooring hospitality. In J. Germann Molz and S. Gibson (eds) *Mobilizing Hospitality. The Ethics of Social Relations in a Mobile World* (pp. 1–26). Aldershot: Ashgate.

Grimwood, B.S.R., Doubleday, N.C., Ljubicic, G.J., Donaldson S.G. and Blangy, S. (2012) Engaged acclimatization: Towards responsible community-based participatory research in Nunavut. *The Canadian Geographer* 56 (2), 211–230.

Haanpää, M., Hakkarainen, M. and Harju-Myllyaho, A. (2018) Katsaus yhteiskunnalliseen yrittäjyyteen matkailussa: osallisuuden mahdollisuudet pohjoisen urbaaneissa paikallisyhteisöissä (Review of social entrepreneurship in tourism: possibilities of participation in the Northern urban communities). *The Finnish Journal of Tourism Research* 14 (2), 44–58.

Hakkarainen, M. (2017) Matkailutyön ehdot syrjäisessä kylässä (The conditions of tourism work in a remote village. PhD thesis. Acta Universitatis Lapponiensis 357, Rovaniemi: Lapland University Press.

Hall, J., Matos, S., Sheehan, L. and Silvestre, B. (2012) Entrepreneurship and innovation at the base of the pyramid: A recipe for inclusive growth or social exclusion? *Journal of Management Studies* 49 (4), 785–812.

Hall, M.C. (2003) Politics and place: An analysis of power in tourism communities. In S. Singh, D.J. Timothy and R.K. Dowling (eds) *Tourism in Destination Communities* (pp. 99–114). Wallingford: CABI Publishing.

Hashimoto, A. (2014) Tourism and socio-cultural development issues. In R. Sharpley and D.J. Telfer (eds) *Tourism and Development: Concepts and Issues* (2nd edn) (pp. 205–236). Bristol: Channel View Publications.

Hickey, S. and Mohan, G. (eds) (2004) *Participation: From Tyranny to Transformation?* New York, NY: Zed Books.
Hinch, T. and Butler, R. (1997) Indigenous tourism: a common ground for discussion. In R. Butler and T. Hinch (eds) *Tourism and Indigenous Peoples* (pp. 3–19). London: International Thomson Publishing.
Hinch, T. and Butler, R. (2009) Indigenous tourism. *Tourism Analysis* 14 (1), 15–27.
Höckert, E. (2011) Community-based tourism in Nicaragua: A socio-cultural perspective. *The Finnish Journal of Tourism Research* 7 (2), 7–25.
Höckert, E. (2018) *Negotiating Hospitality: Ethics of Tourism Development in the Nicaraguan Highlands.* London: Routledge.
Höckert, E., Hakkarainen, M. and Jänis, J. (2013) Matkailun paikallinen kehittäminen maaseudulla. In S. Veijola (ed.) *Matkailututkimuksen lukukirja* (pp. 160–172). Rovaniemi: Lapin Yliopistokustannus.
Jamal, T. (2019) *Justice and Ethics in Tourism.* London: Earthscan.
Jamal, T. and Dredge, D. (2014) Tourism and community development issues. In R. Sharpley and D. Telfer (eds) *Tourism and Development: Concepts and Issues* (2nd edn) (pp. 178–204). Bristol: Channel View Publications.
Jamal, T. and Getz, D. (1995) Collaboration theory and community tourism planning. *Annals of Tourism Research* 22 (1), 186–204.
Jamal, T. and Stronza, A. (2008) "Dwelling" with ecotourism in the Peruvian Amazon: Cultural relationships in local-global spaces. *Tourist Studies* 8 (3), 313–336.
Jamal, T., Everett, J. and Dann, G. (2003) Ecological rationalization and performative resistance in natural area destinations. *Tourist Studies* 3 (2), 143–169.
Kant, I. (1996 [1795]) Toward perpetual peace: a philosophical project. In M.J. Gregor (transl.) *Practical Philosophy: The Cambridge Edition of the Works of Immanuel Kant.* Cambridge: Cambridge University Press.
Kaján, E. (2014) Community perceptions to place attachment and tourism development in Finnish Lapland. *Tourism Geographies* 16 (3), 490–511, doi:10.1080/14616688.2014.941916.
Keen, D. and Tucker, H. (2012) Future spaces of postcolonialism in tourism. In J. Wilson (ed.) *The Routledge Handbook of Tourism Geographies* (pp. 97–102). London: Routledge .
Keogh, B. (1990) Public participation in community tourism planning. *Annals of Tourism Research* 17, 449–465.
Koster, R., Baccar, K. and Lemelin, H. (2012) Moving from research ON, to research WITH and FOR Indigenous communities: A critical reflection on community-based participatory research. *The Canadian Geographer* 56 (2), 195–210, doi:10.1111/j.1541-0064.2012.00428.
Kugapi, O. and de Bernardi, C. (2017) Alkuperäiskansamatkailu (Indigenous tourism). In J. Edelheim and H. Ilola (eds) *Matkailututkimuksen avainkäsitteet (Key Concepts in Tourism Research)* (pp. 70–75). Rovaniemi: Lapland University Press.
Länsman, A.S. (2004) *Väärtisuhteet Lapin Matkailussa. Kulttuurianalyysi suomalaisten ja saamelaisten kohtaamisesta (Host–Guest Relations in Tourism in Sápmi).* Oulu: Kustannus-Puntsi.
Lashley, C. (2017) Introduction. Research on hospitality: the story so far/ways on knowing hospitality. In C. Lashley (ed.) *The Routledge Handbook of Hospitality Studies* (pp. 17–26). Abingdon: Routledge.
Lashley, C. (2000) Towards a theoretical understanding. In C. Lashley and A. Morrison (eds) *In Search of Hospitality: Theoretical Perspectives and Debates* (pp. 1–17). Oxford: Butterworth-Heinemann.
Leal, P.A. (2010) Participation: the ascendancy of a buzzword in the neo-liberal era. In A. Cornwall and D. Eade (eds) *Deconstructing Development Discourse. Buzzwords and Fuzzwords* (pp. 539–548). Oxford: Oxfam GB.

Lee, T.H. and Jan, F.-H. (2019) Can community-based tourism contribute to sustainable development? Evidence from residents' perceptions of the sustainability. *Tourism Management* 70, 368–380, doi:10.1016/j.tourman.2018.09.003.

Levinas, E. (1969) *Totality and Infinity. An Essay of Exteriority*. Pittsburgh, PA: Duquesne University Press.

Lundberg, E. (2015) The level of tourism development and resident attitudes: A comparative case study of coastal destinations. *Scandinavian Journal of Hospitality and Tourism* 15 (3), 266–294, doi:10.1080/15022250.2015.1005335.

Lundin, R. (2016) Project society: Paths and challenges. *Project Management Journal* 47 (4), 7–15.

Lüthje, M. (1998) The impacts of tourism on Saami domicile area from the point of view of carrying capacity and Saami culture of the area. In S. Aho, H. Ilola and J. Järviluoma (eds) *Dynamic Aspects in Tourism Development. Proceedings of the 5th Nordic Symposium on Tourism Research* (Vol. 1 pp. 31–46). Rovaniemi: University of Lapland Press.

Lynch, P., Germann Molz, J., McIntosh, A., Lugosi P. and Lashley, P. (2011) Theorizing hospitality. *Editorial in Hospitality & Society* 1 (1), 3–24.

Mathieson, A. and Wall, G. (1982) *Tourism: Economic, Physical and Social Impacts*. London: Longman.

Miettinen, S. (2007) *Designing the Creative Tourism Experience: A Service Design Process with Namibian Crafts People*. Publication series of University of Art and Design Helsinki A 81. Jyväskylä: Gummerus kirjapaino Oy.

Minghetti, V. and Buhalis, D. (2010) Digital divide in tourism. *Journal of Travel Research* 49 (3), 267–281, doi:10.1177/0047287509346843.

Minnaert, L., Maitland, R. and Miller, G. (2011) What is social tourism? *Current Issues in Tourism* 14 (5), 403–415.

Moscardo, G. (2008) Community capacity building: An emerging challenge for tourism Development. In G. Moscardo (ed.) *Building Community Capacity for Tourism Development* (pp. 1–16). Oxfordshire: CABI.

Müller, D.K. and Huuva, S.K. (2009) Limits to Sami tourism development: The case of Jokkmokk, Sweden. *Journal of Ecotourism* 8 (2), 115–127, doi:10.1080/14724040802696015.

Müller, D.K. and Pettersson, R. (2006) Sámi heritage at the winter festival in Jokkmokk, Sweden. *Scandinavian Journal of Hospitality and Tourism* 6 (1), 54–69, doi:10.1080/15022250600560489.

Müller, D.K. and Viken, A. (2017) Indigenous tourism in the Arctic. In A. Viken and D.K. Müller (eds) *Tourism and Indigeneity in the Arctic* (pp. 3–15). Bristol: Channel View Publications.

Nicholas, L. and Thapa, B. (2018) Tourism in the Fond Gens Libre Indigenous community in Saint Lucia. Examining impacts and empowerment. In I.B. de Lima and V. King (eds) *Tourism and Ethnodevelopment: Inclusion, Empowerment and Self-determination* (pp. 246–259). Abingdon: Routledge.

Niskala, M. and Ridanpää, J. (2016) Ethnic representations and social exclusion: Sáminess in Finnish Lapland tourism promotion. *Scandinavian Journal of Hospitality and Tourism* 16 (4), 375–394, doi:10.1080/15022250.2015.1108862.

NPA (Northern Periphery and Arctic Programme) (2018) Programme manual. See http://www.interreg-npa.eu/for-project-managers/programme-manual/ (accessed May 2019).

O'Gorman, K. (2010) *The Origins of Hospitality and Tourism*. Oxford: Goodfellow Publishers.

Prasetyo, N., Carr, A. and Filep, S. (2019) Indigenous knowledge in marine ecotourism development: The case of *Sasi Laut*, Misool, Indonesia. *Tourism Planning & Development* 17 (1), 46–61, doi:10.1080/21568316.2019.1604424.

Pyyhtinen, O. (2014) *The Gift and its Paradoxes. Beyond Mauss*. Farnham: Ashgate.

Rantala, K. and Sulkunen, P. (2006a) Esipuhe (Foreword). In K. Rantala and P. Sulkunen (eds) *Projektiyhteiskunnan kääntöpuolia (Downsides of Project-Society)* (pp. 7–14). Helsinki: Gaudeamus.

Rantala, K. and Sulkunen, P. (2006b) Uudet hallintamekanismit – miksi ja miten? (New mechanism for management – how and why?) In K. Rantala and P. Sulkunen (eds) *Projektiyhteiskunnan kääntöpuolia (Downsides of Project-Society)* (pp. 15–16). Helsinki: Gaudeamus.

Ren, C. and Jóhannesson, G.T. (2018) Collaborative becoming: Exploring Tourism Knowledge Collectives. In C. Ren, G.T. Jóhannesson and R. van der Duim (eds) *Co-creating Tourism Research: Towards Collaborative Ways of Knowing* (pp. 24–38). London: Routledge.

Ren, C., Jóhannesson, G.T. and van der Duim, R. (2018) Towards a collaborative manifesto: Configurations of Tourism knowledge co-creation. In C. Ren, G.T. Jóhannesson and R. van der Duim (eds) *Co-creating Tourism Research: Towards Collaborative Ways of Knowing* (pp. 179–183). London: Routledge.

Saarinen, J. (2006) Traditions of sustainability in tourism studies. *Annals of Tourism Research* 33 (4), 1121–1140.

Saarinen, J. (2010) Local tourism awareness: Community views in Katutura and King Nehale Conservancy, Namibia. *Development Southern Africa* 27 (5), 713–724.

Sámi Parliament (2018) Vastuullisen ja eettisesti kestävän saamelaismatkailun toimintaperiaatteet (Principles of responsible and ethically sustainable Sámi tourism). See https://dokumentit.solinum.fi/samediggi/?f=Dokumenttipankki%2FKertomukset%2C%20ohjelmat%20ja%20suunnitelmat (accessed May 2019).

Sammels, C.A. (2014) Bargaining under thatch roofs: Tourism and the allure of poverty in highland Bolivia. In D. Picard and M.A. Giovine (eds) *Tourism and the Power of Otherness: Seductions of Difference* (pp. 124–140). Bristol: Channel View Publications.

Scheyvens, R. (1999) Ecotourism and the empowerment of local communities. *Tourism Management* 20, 245–249, doi:10.1016/S0261-5177(98)00069-7.

Scheyvens, R. (2002) *Tourism for Development: Empowering Communities*. Harlow: Pearson Education.

Scheyvens, R. (2003) Local involvement in managing tourism. In S. Singh, D.J. Timothy and R.K. Dowling (eds) *Tourism in Destination Communities* (pp. 229–252). Wallingford: CABI Publishing.

Scheyvens, R. (2011) *Tourism and Poverty*. London: Routledge.

Scheyvens, R. and Biddulph, R. (2017) Inclusive tourism development. *Tourism Geographies* 20 (4), 589–609, doi:10.1080/14616688.2017.1381985.

Schilcher, D. (2007) Growth versus equity: The continuum of pro-poor tourism and neoliberal governance. In M. Hall (ed.) *Pro-poor Tourism: Who Benefits? Perspectives on Tourism and Poverty Reduction* (pp. 57–83). Clevedon: Channel View Publications.

Smith, V.L. (ed.) (1977) *Hosts and Guests: The Anthropology of Tourism*. Philadelphia, PA: University of Pennsylvania Press.

Stiefel, M. and Wolfe, M. (1994) The many faces of participation. In A. Cornwall (ed.) *The Participation Reader*. London: Zed Books.

Sulkunen, P. (2006) Projektiyhteiskunta ja uusi yhteiskuntasopimus (Project society and the new societal contract). In K. Rantala and P. Sulkunen (eds) *Projektiyhteiskunnan kääntöpuolia (Downsides of Project-Society)* (pp. 17–38). Helsinki: Gaudeamus.

Swarbrooke, J. (2002) *Sustainable Tourism Management*. Wallingford: CABI Publishing.

Tanga, P.T. and Maliehe, L. (2011) An analysis of community participation in handicraft projects in Lesotho. *Anthropologist* 13 (3), 201–210.

Telfer, D.J. (2009) Development studies and tourism. In T. Jamal and M. Robinson (eds) *The SAGE Handbook of Tourism Studies* (pp. 504–520). Los Angeles, CA: SAGE.

Telfer, D.J. and Sharpley, R. (eds) (2008) *Tourism and Development in the Developing World*. Abingdon: Routledge.

Telfer, E. (2000) The philosophy of hospitableness. In C. Lashley and A. Morrison (eds) *In Search of Hospitality: Theoretical Perspectives and Debates* (pp. 38–55). Oxford: Butterworth-Heinemann.

Tosun, C. (2000) Limits to community participation in the tourism development process in developing countries. *Tourism Management* 21 (6), 613–633.

Tucker, H. (2014) Mind the gap: Opening up spaces of multiple moralities in tourism encounters. In M. Mostafanezhad and K. Hannam (eds) *Moral Encounters in Tourism* (pp. 199–208). Farnham: Ashgate.

Tuulentie, S. (2006) The dialectic of identities in the field of tourism. The discourses of the Indigenous Sámi in defining their own and the tourists' identities. *Scandinavian Journal of Hospitality and Tourism* 6 (1), 25–36.

Tuulentie, S. and Sarkki, S. (2009) Kun kylästä tulee keskus (When village turns into tourism destination). In S. Tuulentie (ed.) *Turisti tulee kylään. Matkailukeskukset ja lappilainen arki (Tourists as Guests in Finnish Tourism Destinations)* (pp. 9–28). Helsinki: Minerva Kustannys Oy.

van der Duim, R., Peters, K. and Akama, J. (2006) Cultural tourism in African communities: A comparison between cultural Manyattas in Kenya and cultural tourism project in Tanzania. In M.K. Smith and M. Robinson (eds) *Cultural Tourism in a Changing World: Politics, Participation and (Re)presentation*. Clevedon: Channel View Publications.

Veijola, S. and Jokinen, E. (2008) Towards a hostessing society? Mobile arrangements of gender and labour. *NORA: Nordic Journal of Feminist and Gender Research* 16 (3), 166–181.

Veijola, S., Germann Molz, J., Pyyhtinen, O., Höckert, E. and Grit, A. (2014) *Disruptive Tourism and its Untidy Guests. Alternative Ontologies for Future Hospitalities*. New York: Palgrave MacMillan.

Viken, A. and Müller, D.K. (2017) *Tourism and Indigeneity in the Arctic*. Bristol: Channel View Publications.

Warnholz, G. and Barkin, D. (2018) Development for whom? Tourism used as a social intervention for the development of indigenous/rural communities in natural protected areas. In I.B. de Lima and V. King (eds) *Tourism and Ethnodevelopment: Inclusion, Empowerment and Self-determination* (pp. 27–43). Abingdon: Routledge.

Wearing, S. and McDonald, M. (2002) The development of community-based tourism: Re-thinking the relationships between tour operators and development agents as intermediaries in rural and isolated area communities. *Journal of Sustainable Tourism* 10 (2), 31–45.

Wearing, S. and Wearing, M. (2014) On decommodifying ecotourism's social value: Neoliberal reformism or the new environmental morality? In M. Mostafanezhad and K. Hannam (eds) *Moral Encounters in Tourism* (pp. 123–138). Farnham: Ashgate.

Zapata, M.J., Hall, C.M., Lindo, P. and Vanderschaeghe, M. (2011) Can community-based tourism contribute to development and poverty alleviation? Lessons from Nicaragua. *Current Issues in Tourism* 14 (8), 725–749.

2 Resourcing the Self: Forms of Capital and Stratification in the Platform Hospitality Community

Andreja Trdina, Salla Jutila and Maja Turnšek

Introduction

The platform economy has generated immense excitement with its promise of transforming work and consumption through technology and novel socioeconomic relations (Fitzmaurice *et al.*, 2018). Yet, as argued by Jansson and Lindell (2018: 1), in contemporary times characterised by connective and locative media, social space and media become even more closely interwoven, making social power relations more fluid and at the same time more technologically dependent. In this chapter, we address the critical issues of the platform economy trend – more precisely, the barriers to participate in it (despite its formal claims of inclusivity and openness) and, as a result, the reproduction of social inequalities. Taking the Airbnb global hospitality community as a case study and through the perspective of critical hospitality studies, we aim to examine the principles of inequality that underpin the platform economy.

Broadly, this chapter addresses the changing modes of social reproduction that characterise our technologically mediated society. Our analysis relates the users of technology to their social conditioning. It is rather obvious that material resources are crucial for participating in the Airbnb platform and for successful hosting: one has to have a property to offer to be able to gain social advantages in the platform economy. This already reflects the principal dividing line between 'the haves' and 'the have-nots'. However, from a social inequalities perspective, it is important to discern that, besides economic capital, access to the platform economy also requires other forms of capital. Hence we have deliberately reverted our focus to the as-yet ignored issue of the required cultural competences to successfully participate in and navigate the Airbnb platform to one's advantage. It is our contention that the social structure has not been sufficiently problematised in current debates on mediatised hospitality, which often fail to engage in the discussion of power relations and ignore 'the socially structured and structuring role of media technologies and practices' (Jansson & Lindell, 2018: 1). We thus call for a sociological approach to platform technologies and to the discourses and uses they promote. We aim to grasp the concept of platform stratification by focusing particularly on the structurally unequal distribution of required cultural resources, assets and skills among service providers and hosts (see Haanpää et al., 2018: 52), examining new divisions and strategies of distinction among the hosts themselves.

Our contribution is grounded in Skeggs' (2004) broader argument about the ways in which class inequalities are reshaped and reproduced through the privileges of individualisation in today's society. Within this context, we would like to argue that participating in the platform economy is not only limited to the distribution of conventional material resources that one could be offering online, but also to the distribution of cultural, emotional and aesthetic resources that structure the online performances of self and the presentations of one's listings. Turnšek and Ladkin (2017) point out that in traditional online travel agencies (e.g. Booking.com, Trivago), the accommodation providers at first presented themselves as professional organisations, not individuals. Airbnb, on the other hand, riding on the wave of the popular sharing discourse popularised by Couchsurfing, was among the first that expected both guests and hosts to present themselves as individuals, with profile pictures and descriptions. What we are seeing nowadays is that differences in the practices on different platforms are not as strict and clear any longer: for example, today it is possible to have a company profile on Airbnb and an individual profile on Booking.com. However, our focus here is on the aspects of the platforms where the presentation is focused on individuals, not organisations. Airbnb called this one of the key dimensions of design for building trust among strangers. These key 'trust' mechanisms have recently been researched, with a primary focus on their effects on guests, including the mechanisms of the review features (Hakkarainen & Jutila, 2017; Teubner et al., 2016; Zervas et al., 2015), hosts' pictures (Eyal et al., 2016; Fagerstrøm et al., 2017) and Airbnb's advertisements (Liu & Mattila, 2017). Not much is known, however, on how these mechanisms interrelate with unequally distributed cultural, emotional and aesthetic resources that structure online performances of self.

As Skeggs argues 'The ability to propertize one's self and one's culture (as an exchange value) generates new forms of exploitation based on immateriality' (Skeggs, 2004: 176). This is well reflected in the principles of the sharing economy as a new business model that personalises market

exchange and monetises previously non-commodified realms of (private) life. Therefore, beyond the incontestable significance of material resources as the evident condition for inclusion in the platform economy, the present study examines the ways in which one's cultural dispositions and ways of existing in and interacting with the world fuel the privileges and social divisions in hospitality platforms. In this way, we highlight how culture functions as capital and how immaterial aspects are being put to work to gain social advantages in a digital, algorithmic world.

To this end, we first introduce the transformations of global social inequalities in the context of the new platform economy, reflecting on what kinds of challenges platforms pose to social organisation in general. After deconstructing the myth of meritocracy, we proceed to elaborate briefly on how platforms discipline hosts through entrepreneurship ethos and meritocratic discourse. In what follows, we outline the demand for emotional and aesthetic skills in hospitality platforms in the context of social inequalities, maintaining that platforms legitimate these resources as desirable dispositions while ignoring the fact that access to these resources is rendered structurally unequal. Finally, in order to examine how emotional and aesthetic labour is performed and how emotional and aesthetic capital operate as a decisive currency in the online hospitality network, we consider qualitative content analysis of online discussions and online profiles of Finnish and Slovenian Airbnb hosts. Both societies share certain cultural specifics as regards social stratification; for example, they are both fairly egalitarian societies (reflected in lower Gini coefficients from the EU average in 2017, according to Eurostat), characterised predominantly by the ideology of egalitarianism. We argue that hospitality platforms – or 'technologies of hospitality' as they were insightfully termed by Bialski (2012) – are recontextualising and reinforcing existing social divisions. Our theoretical stance is grounded in Bourdieu's fundamental argument about culture as symbolic distinctions that are implicated in maintaining social inequalities (Bourdieu, 2010) and Skeggs' argument on how today's inequalities are reproduced primarily through cultural aspects and privileges of individualisation (Skeggs, 2014).

Global Social Inequalities and Platform Work

According to Jansson (2016: 422–423), the globalisation process is transforming class structures via two trends. The first is the emergence of new influential class fractions with a transnational character, reflected both in their expanding symbolic competences and extended modes of mobility and interaction. The second is the trend of the conditions of older (industrial and political) elite groups being challenged by instant communication and global networking, which demand higher degrees of reflexivity, mobility and networking. Under such conditions, the significance of cosmopolitan capital is growing, taking on a key role in social stratification (Igarashi & Saito, 2014). Jansson defines cosmopolitan capital as referring to 'embodied capabilities and attributes, such as language skills, international experiences and educational degrees, that make social agents better equipped to navigate the world geographically, socially and culturally' (Jansson, 2016: 423).

With the platform economy we are witnessing a new reality in which virtual and non-virtual work blend, transgressing the dichotomy altogether. Platforms operate via algorithms, computations that define their operations; for example, platforms' search engines rank hosts according to specific algorithms that are usually proprietary and secret. According to Striphas (2015: 406), algorithms have significantly taken on what has been primarily a culture's responsibility, namely the task of reassembling the social sphere. Within what Striphas insightfully terms algorithmic culture, algorithms are ever more decisive and 'culture is fast becoming – in domains ranging from retail to rentals, search to social networking, and well beyond – the positive remainder resulting from specific information processing tasks, especially as they relate to the informatics of crowds' (Striphas, 2015: 406). What is at stake in such algorithmic intervention, Striphas argues, is privatisation of the process of decision-making and the contestation that comprises the ongoing struggle to determine the cultural values, practices and artefacts of social groups.

In her contribution on moral economy in the context of the alternative tourism of the non-commodified Couchsurfing community, Germann Molz (2013) convincingly demonstrates the way technology and algorithms also shape social relations in networked hospitality. By demonstrating how engaging in caring relationships is enabled by various social networking mechanisms (among others), Germann Molz (2013) encourages us to frame the issue of participation in platforms within the questions of controlling, regulating and disciplining the subjects. Similarly, Roelofsen and Minca (2018) reflect on the biopolitics that are inherent to the quantifications and qualifications produced and established by the global Airbnb community. They maintain that the metrics (deemed to be objective) are supposed to represent a certain 'veracity' about a person's or place's authenticity and performance, while in fact their procedures incidentally create hierarchies among users, providing incentives and optimising their performance by increasing their visibility and thus their value in the Airbnb community (Roelofsen & Minca, 2018: 178).

Furthermore, current discussions indicate that entrepreneurship opportunities in the platform economy occur in a socially selective way. Schor (2017) claims that platform capitalism recontextualises and reinforces existing social divisions because it is increasing income inequality among the bottom 80% of the income distribution, for primarily two reasons. First, many providers have full-time jobs and they use platforms to augment their incomes. The work that people are taking on appears to be new, not a substitute for other income-earning opportunities. Second, most providers are highly educated and they are taking on tasks that have traditionally been done by workers of low educational attainment, such as cleaning, driving and other manual labours, suggesting a crowding-out effect. Similarly, a study by the JPMorgan Chase Institute (2016) identified a stark difference between work on 'capital' platforms (such as Airbnb or the car-sharing site Turo) and what they define as 'labour' platforms (e.g. Uber or TaskRabbit): people providing services via 'capital' platforms are bringing in supplemental income, while people in the 'labour' platforms typically work to offset shortfalls in their monthly earnings. It is also important to consider how working and entrepreneurship possibilities appear geographically: rural vs. urban and city centers vs. urban peripheries (see Haanpää et al., 2018: 52).

The platform economy, whilst representing a fundamental shift in the nature of work and the relationship between labour and capital, poses a relevant and timely challenge to social organisation in general. There is a need to address the impact of the platform economy on transformations of (global) class relations and dig deeper into platform uses and their social embeddedness. We are thus interested in the formations of new divisions that are established and maintained within the platform context, how these are reflected in ever more expanding cultural competences and the strategies of providers required for successful participation in the platform economy. On the basis of our analysis, we aim to discuss the broader issue of recontextualising and transforming existing (global) social inequalities. Before we delve into this in greater depth, we wish to problematise the ideology of meritocracy as embedded in the workings of contemporary inequalities (in the sense that everyone has an equal opportunity for social mobility) and briefly present how platforms employ meritocratic discourses in promoting the idea of the entrepreneurial self.

Entrepreneurship Ethos: Disciplining (Super)hosts through Meritocratic Logic

Today, meritocracy – grounded in the idea of equal opportunity and social mobility – represents a key cultural means for legitimating social differences. While promising equal opportunities, it in fact creates and legitimises social divisions. Littler (2018) argues that meritocratic discourse effectively absorbs the language of equality into the idea of entrepreneurial self-fashioning and self-realisation. Therefore, meritocracy is, according to Littler (2018: 15), a formative, yet under-theorised ideological engine of late capitalism: 'Neoliberal meritocracy, as a potent blend of an essentialised notion "talent", competitive individualism and belief in social mobility, is mobilized to both disguise and gain consent for the economic inequalities' (Littler, 2018: 223).

Through the rise of meritocracy discourse, we can better understand the workings of contemporary inequality. As noted by Khan (2011: 9), the concept of talent-based 'merit' has replaced old notions of social status with personal attributes such as hard work, discipline and other forms of human capital that can be evaluated separately from one's social life conditions. Khan (2011: 9) further argues that 'meritocracy of hard work and achievement has naturalized socially constituted distinctions, making differences in outcomes appear a product of who people are rather than a product of the conditions of their making'. Similarly, Mijs (2018) argues that inequality is produced by the ways in which talent is defined, institutionalised and sustained by the moral deservingness we attribute to the accomplishments of talents. Consequently, today's inequalities are as striking as ever, yet harder to challenge than ever before. Mijs (2018) notes that talent is a major means through which people seek rent in modern-day capitalism. According to Mijs (2018), the return on talent relies on three processes, none of which is in itself meritocratic.

(1) Just like capital, talents are inter-generationally transmitted from parents to children.
(2) What constitutes talent relies on the definitional power of powerful gatekeepers who decide which traits to reward and which to discard.
(3) Talents are structural, not individual traits – their economic returns are institutionalised in privileged positions.

The neoliberal meritocratic ideal, as convincingly argued by Littler (2018: 3), in this way conveniently disregards systematic inequality, social location and the head start accrued by the children of those at the top of social ladder, hence ignoring the fact that climbing up the social hierarchy is not equally easy for everyone or, in other words, it is simply more challenging for some than for others.

In what follows, we briefly describe the logic of Airbnb and how the platform employs meritocratic discourses while endorsing the idea of entrepreneurialism. Airbnb attracts people to become a host and list their homes (or apartments they own) on Airbnb by marketing that 'Listing a home on Airbnb has never been easier or more customizable'. Suspicions of potential hosts are intended to be assuaged by corporate statements that over 2 million people host on Airbnb and that no host's style is the same. It is portraying an image that everything is controlled by the hosts themselves: hosts can decide when they host, set a price they feel good about, establish rules for their own space and connect their own calendars with the Airbnb calendar to avoid overlapping schedules (Airbnb, 2019a).

While giving all control and freedom to hosts, Airbnb also sets a number of requirements for them. A host has to maintain a kind of booking system in Airbnb; that is, they block off dates when the apartment is not available, plan minimum and maximum nights a guest can stay, as well as how far in the future guests can book. Hosts have to master at least basic knowledge in pricing as they need to choose nightly prices. Even though there is a pricing tool to help with this, hosts are also asked to add custom details, such as cleaning fees, weekly discounts and special prices for specific times of year. This requires at least some understanding about profitability and the market-driven system. To list an apartment on

Airbnb, a personal profile is also required. To set up a profile, a photo, personal contact details, bank account details and a photo of a government ID are required. In some cases, an additional photo of the host different from the profile picture is also required. Implementing all this requires organising skills, sovereignty in the use of technology and – above all – competence in communication.

Platform logic and all their requirements inevitably – although imperceptibly – affect and shape the way hosts think about and position themselves, predominantly refashioning themselves as entrepreneurs, establishing their own business and building what are now seemingly indispensable entrepreneur selves. For example, pricing makes hosts value their own properties and compare them with other properties, with competition being the organising principle. This is what happens through Airbnb's logic and requirements that vigorously incite what Littler (2018: 172) would term 'internalised meritocratic subjects'. With the promise of open opportunity, the hospitality platform endorses entrepreneurialism and competitive individualism among providers, and in addition valorises particular individual skills and traits. In this sense, Airbnb is constituting its own talents and these talents are later rewarded through platform algorithm logic. Airbnb thus operates here as a powerful 'gatekeeper', deciding which traits to reward and which to discard (see Mijs, 2018; Striphas, 2015: 406). Defined talents socialise hosts into the system of global networking, which demands embodied capabilities and attributes such as higher degrees of responsiveness, flexibility and networking. In this system, the economic returns of the talents are derived by those in privileged position, as the algorithms rank listings based on the talents of hosts. Listings of hosts with the best talents appear first on the list when searching by city name.

An illustrative example of the reward system is the superhost programme on Airbnb. Hosts meeting certain performance standards – as measured by an automated system – qualify for superhost status. Once a host reaches superhost status, a badge indicating this will automatically appear on their listing and profile. The requirements for superhosting are based on the number and length of reservations, maintained review and response rates, scarcity of cancellations and overall ratings from guests. Airbnb describes superhosts as 'experienced hosts who provide a shining example for other hosts and extraordinary experiences for their guests'. A superhost's activity is monitored a couple of times per year to ensure that the programme highlights the most dedicated hosts providing outstanding hospitality (Airbnb, 2019b). The superhost – the champion of Airbnb hospitality – is thus constructed as a biopolitical horizon; as Roelofsen and Minca (2018: 170) argue, it is 'the incarnation, identified via the algorithms of that specific platform, of all the qualities requested to succeed and emerge in Airbnb's global community of hospitality'. In this sense, we consider the travel platform not only as 'a complex socio-technical system', binding an individual level of social activities with technological operations that interact with users – and a structural level of social (trust) and technological phenomena (control, oversight), as argued by Rasmussen (2014: 117) – but also as a socio-technical system that, by its own systemic processes and mechanisms of reviewing and ranking, establishes different statuses of providers and in this way forms a distinct stratification order of the platform community. In the following section, we aim to address the role of emotional and aesthetic resources in platform stratification order and ignite the discussion on their relatedness to social inequalities in general.

Emotional and Aesthetic Capital as Key Currency in Online Hospitality

We believe the attention to cultural, emotional and aesthetic resources brings a much-needed perspective to understanding the often less visible, yet not less salient, aspects of symbolic distinction that are deeply implicated in social inequalities. Such a perspective, we contend, offers a better footing on which to base the study of stratification in the era of algorithmic culture and the platform economy. From this perspective, Rasmussen (2014) offers a promising account on capital dynamics in social media, highlighting the role of skills in one's so-called 'networked lifeworld'. Rasmussen's contribution concerns the role of social capital in particular and calls to identify 'not only the social implications of social networks, but also the necessary social conditions among users that stimulate social capital accumulation' (Rasmussen, 2014: 125).

Following on from this, it is our interest to explore the ways in which the mastering of digital media and hospitality platforms is structured along particular skills produced in and through one's networked lifeworld and how these play a prominent differentiating role between platform users as regards accumulating and retaining advantages. We want to call attention to emotional and aesthetic capital in particular as the key currency in hospitality platforms. We contend that those with access to emotional and aesthetic capital can expect returns that exceed their initial investments in the given capital. Following the work of Savage *et al.* (2005), we understand emotional and aesthetic resources as capital, not in terms of distinct relations of exploitation, but from the perspective of its potential to be accumulated and converted to other resources over time and in certain social, cultural and institutional settings. According to Rasmussen (2014), skills transform resources of networks into capital, implying that skills and capital are interdependent properties of the social network. It is 'the interplay between individual skills and network' that builds the capital (Rasmussen, 2014: 114).

Our contribution is grounded in Skeggs' (2004) broader argument about the ways in which class inequalities in today's society are reshaped and reproduced through the privileges of individualisation. Similarly, Bottero (2004) recognises the failure of collective class identities and argues for a new direction in class analysis, abandoning the notion of distinct class identities or groups and focusing rather on 'class as an individualized process of hierarchical distinction' (Bottero, 2004: 991). Following Bourdieu's (2010) legacy, class now has to be discovered in the ways in which individuals differentiate themselves through cultural tastes and practices, so that the emphasis of such renewed class analysis (also called the 'cultural class approach') should be on how specific cultural practices are bound up in the reproduction of social hierarchies and inequality. Bottero (2004: 989) maintains that class processes operate through individualised distinctions rather than social groupings. Instead of collective class identity, the classed nature of particular cultural practices (at the individual level) is at the heart of understanding contemporary class relations and how inequality works today. In this sense, class processes have become more implicit, yet their effects are no less pervasive. Our main argument is thus that participating in the platform economy is not only limited to the distribution of conventional material resources, but also to the structurally unequal distribution of cultural (emotional and aesthetic in particular) resources that structure online performances of self and of hospitality acts and, in turn, reinforce social inequalities.

According to Turnšek and Ladkin (2017), portraying a smiling, trustworthy image of accommodation hosts, combined with intensive social-media-like communication with guests, may extend the emotional labour of the host (Hochschild, 1983; Veijola *et al.*, 2013). They argue that inherent in such self-promotion is also the extension of aesthetic labour – the employee's capacity to 'look good' (Warhurst *et al.*, 2000) and the aesthetic efforts made to manage the impressions of the apartment offered by staging and selection of photos. Yet, as others have pointed out (Baum, 2007), the emergence of aesthetic labour poses a serious concern relating to the appropriate management of diversity in the tourism workplace. Encounters characterised by care, intimacy and authenticity are also more likely to be valorised in the platform community. In the case of the Airbnb platform, Roelofsen and Minca (2018) critically reflect on how hospitality translates into a metrics of care as well as reflecting on localness and on how specific practices related to the 'spatialities of the home' are central to the qualification/quantification of life and of living spaces generated by the platform. Roelofsen and Minca (2018: 170) claim that platforms like Airbnb have in fact turned the labour of care of others in the private sphere into exchange value. Guests are – by platforms in particular – encouraged to assess their hosts along a set of hosting standards. Although some of these standards include several material and locational aspects of the home, many also refer to a subjective set of affective practices and embodied performances on the part of the host, as noted by Roelofsen and Minca (2018: 176) others, such as being communicative, accurate in the description of a place, and attentive and considerate in responding to requests. According to Roelofsen and Minca (2018), Airbnb also suggests that hosts may receive an optimal rating through performances that make guests 'feel welcome' and 'at home', for instance by 'making small gestures', 'avoiding confusion' and 'personalising each guest's experience'. Other rated elements (e.g. the check-in process) equally imply a set of embodied practices that expects hosts to be apt, available, clear, attentive and authentic. Hosts are accordingly impelled to qualify as optimally communicative and hospitable primarily through their emotional and aesthetic labour.

All these characteristics are drawn into systems of measurement and given value by Airbnb. Thus, providers face barriers to entry and successfully participation in the platform economy depending not only on their material assets, but also their cultural and symbolic skills. These include, for example, aesthetic skills (professional or high-definition photos of styled apartments) and the ability to digitally represent their home (both visually and textually) in a market-appealing manner, strong foreign language skills, digital literacy, and knowledge about and manipulation of algorithms (Lee *et al.*, 2015) and rating improvement strategies. According to Schmid-Drüner (2016), the rating and reputation mechanisms of platforms may lead to an increase in 'begging and bragging rituals' to land the next job. De Groen *et al.* (2016) discuss how it is difficult to identify the content of algorithms and thus to pinpoint how important users' ratings exactly are in the oversight of workers. However, on ListMinut (a Belgian personal services platform), profiles with good ratings were found to be typically awarded a disproportionate amount of jobs – the 6% of earners with the highest ratings earn about one-third of the total revenues on the platform. De Groen *et al.* (2016) conclude that, in addition to the standard determinants of workers' earnings (e.g. gender, age, occupation etc.), the characteristics and evaluation mechanism of the platform have a significant influence on the distribution of tasks and earnings.

For our discussion, we thus understand emotional and aesthetic capital in Bourdieusian terms as field-specific cultural capital in the platform economy – that is, resources, qualities and traits that pay off in a particular (structural) setting, meaning that they have potential to be accumulated and converted to other resources. According to Savage *et al.* (2005: 45), for a resource to become a form of capital, there should be systemic processes allowing the garnering of such resources by those who possess them. This could certainly be argued for hospitality platforms that offer situations where advantages accumulate over time. The sharing of emotional resources that underpins the so-called 'moral economy' of network hospitality (Germann Molz, 2013) is afforded by many aspects of the platform's profile template and technical mechanisms, which produce the conditions of the acts in which hospitality hosts and guests engage. Furthermore, we maintain that platforms legitimate these resources as desirable dispositions, yet access to these resources is rendered structurally unequal. They are unequally distributed among providers who capitalise on them (by transforming them into symbolic and economic capital). Therefore, in this new digital context, they form a specific stratification order, which we wish to discuss in more detail. This is the main point of interest we take up in our analysis, but first we describe the methodological approach of our study.

Qualitative Content Analysis of Online Community Discussion and Online Self-Presentation

To illustrate the arguments presented in the theoretical framework, we carried out an explorative analysis of online discussions and online profiles of Finnish and Slovenian Airbnb hosts. Both societies share certain cultural specifics with regard to social stratification; for example, both are characterised predominantly by egalitarianism. The two societies also undeniably have many differences regarding history and actualisation of the welfare state, which affect – among other things – the responding to and understanding of social inequalities. There are also certain differences in terms of regulations and requirements regarding Airbnb. However, in this research, Finland and Slovenia perform as geographical sample areas of a global phenomenon and analysis about the societal backgrounds of the countries is not deeply considered.

The analysis consists of two parts. First, we analysed conversations on the Airbnb community, which is where Airbnb hosts can connect and discuss with other hosts. By qualitative content analysis of community discussions, we aimed to address the efforts and tactics of Airbnb providers to successfully participate in platforms. Second, to reinforce our arguments based on the online discussions, we also analysed profiles, listings and the online presence of Finnish and Slovenian Airbnb hosts. We limited our sampling to the capitals of the countries, Helsinki and Ljubljana. By analysing listings and profiles, we aimed to provide an exploratory assessment of new divisions and strategies of distinction among platform providers.

For analysis of the community discussions, we searched discussions relevant to us using the words 'Finland' (68 discussions) and 'Slovenia' (34 discussions). These conversations were either started by a host living in Finland/Slovenia or involved a host living in one of the two countries joining another conversation but mentioning Finland/Slovenia in their discussion. There were also a couple of discussions in which one of the countries was mentioned as an example of something, but such discussions were omitted from our sample. Thus, we analysed the discussions that Finnish and Slovenian hosts started and joined in the community. Through qualitative content analysis of community discussions, we aimed to understand the efforts made in cultivating the required traits and skills, as revealed in the tips and tactics of hosts on how to successfully participate in Airbnb and separate oneself from other hosts.

For the exploratory analysis of hosts' self-presentations, we purposefully selected the profiles provided by Airbnb with the search query 'Ljubljana' and 'Helsinki' – the capitals of Slovenia and Finland. The capital cities were selected as search queries because we assumed that these would be the destinations within the two countries that are, on the one hand, similar in terms of being highly popular to tourists and, on the other, presenting the same type of tourism: urban tourism (while also largely including sports and nature tourism). The selection was limited to the results provided within the first main page after the search query 'Explore', with no definite dates or types of service provision. In this way, we purposefully analysed those hosts' profiles that Airbnb's search engine algorithm rewards by positioning them in the most prominent position. The search query algorithm is a proprietary business secret, yet it also included, among others, the researchers' previous Airbnb use. We checked for the influence of this by comparing two searches made by two researchers from different locations (Finland and Slovenia). The comparison showed that 90% of the listings were the same in the query for Ljubljana and 81% were the same in the query for Helsinki. We included only those listings that were positioned in both queries, again assuming that these were the listings provided with the highest level of salience by Airbnb's algorithm, since they were less affected by the researcher's personal search history and location. Additionally, out of the all profiles published, we excluded those that were repetitions (20% in the case of Ljubljana and 32% in the case of Helsinki, thus analysing each case only once). We excluded restaurant services, since restaurant profiles are made by users and not restaurant providers and because we wanted to focus on how individual providers present themselves, rather than organisations. This left 41 listings and the accompanying providers' profiles for analysis in the case of Ljubljana and 37 in the case of Helsinki, covering both accommodation and experiences. The analysis included the description of the listing and the personal profile's description of the service provider, including accompanying visual material. In short, we thus considered the most prominent profiles as we maintain that these embody all the qualities requested in Airbnb's global community and are therefore rewarded with the highest status in the community. We tried to elaborate on the distinctive characteristics of self-representation in these profiles with regard to emotional and aesthetic labour in particular.

Platform Stratification Order: New Divisions and Strategies of Distinction Among Service Providers

Tactics of hosts: Analysis of online community discussions

By reflecting on the providers' calculative rationalities and tactics that underlie the process of becoming and being a (super)host, we wish to unravel the distinct efforts of providers and how these are qualified and quantified through the platform system. Our analysis of community discussions indicated four themes that recur on the community's online forum: (1) networking, (2) operation management skills, (3) discretion and (4), as a combination of the three previous themes, gaining and keeping the superhost status. These can be seen as the embodied capabilities and attributes that make social agents better equipped to navigate the world of online hospitality platforms (see Jansson, 2016: 423) or as distinct talents that are defined, institutionalised and sustained (see Mijs, 2018) by a platform and are differentiating hosts.

Based on our analysis, it can be argued that networking skills are a crucial talent to be able to succeed in Airbnb hosting. The Airbnb community itself is an illustrative example of this. First, a host has to be aware that this kind of platform for discussions among hosts exists in order to make use of such a networking opportunity. Second, a host needs communication skills in such platforms as they have to internalise a range of habits and informal and unwritten rules for opening a discussion, as well as the ways of presenting questions, suggestions and comments. It is thus compulsory to master how to establish and continue an online discussion with the knowledge of etiquette and the pragmatics of the medium itself. One's digital media literacy, which includes the writing and interpretative skills necessary to communicate effectively via online media, is therefore crucial. Discussion openings mainly start with nice, very brief, but informative presentations about the host and their listing. Comments or questions are often addressed to 'fellow hosts' or even to 'friends'. In general, the networking skills encompass the ability to present oneself to other hosts. The online performance of self in the community can affect the number and quality of the answers a host receives to their comment or question, which again may have effects on the fluency, scope and profitability of the host's accommodation activities.

Networking seems to occur on different levels. In addition to indirect and somewhat spontaneous networking as a by-product of posting questions or comments to the community described above, there are also more direct and intentional networking suggestions presented in community discussions. Airbnb community admins also take part in discussions and, for their part, promote and support networking among hosts. This is done, for example, by suggesting that superhosts create their own community inside the Airbnb community: 'Introduce yourself to other Superhosts below and feel free to share a bit about your home, your region, or your favorite hosting tip. Who knows? Maybe you'll even make some new friends. We're so happy to welcome you as one of our most valued hosts on Airbnb'. Sharing information, arguments and ideas from the central position of a network or – to put it differently – having privileged access to the community discussions indicates the actual operation of social capital as a social mechanism that impacts the exclusions of others when it comes to information and status.

What's more, many discussions contain different kinds of ideas and suggestions about networking among hosts. Hosts are, for example, asking about accommodation for themselves from other hosts and one host also suggested home-swapping among the hosts. In this example, a Slovenian host first joined the discussion supporting the presented idea and later started a chat (within the same discussion) with a host from the Dominican Republic, with both welcoming one another to their beautiful countries as well as promising tips and help. One host, in turn, suggested boosting one another's Airbnb listings on social media: 'I'm going to take a quick look at your pictures to make sure they look cool and then I'm going to add your property to my airbnb favorites'. This is an example of how networking among users and, based on that networking, help in the form of favours are used as a tactic in a sense of what Rasmussen (2014: 57), echoing de Certeau (1984), terms 'the response of dominated subjects, their operating within the processes of dominating system' or, in our case, a platform with all its inherent requirements. Such tactics are to be considered as innovative moderations of established platform routines and therefore as skilful manoeuvres within the game in which the rules are set by other powers.

General operation management skills are another strongly present talent in Airbnb community discussions. Many hosts ask for advice on technical issues in terms of platform management, such as difficulties when trying to publish another listing with the same profile, questions regarding money transfers and confusion about changing functions when using different browsers or languages. Many discussions relate to questions of pricing. Hosts have to master at least basic knowledge of pricing, but they also need to understand the platform's pricing logic. There are many discussions that address smart pricing possibilities and the challenges in using them. Smart pricing is a system that blocks the final price with service fees and cleaning fees included, thereby showing the user only the price set by the host. One host, for example, wondered why her guests get shown different prices for the same period of time: other hosts explained to her that she is probably using smart pricing and advised her to turn it off. Smart pricing is a system that requires both management skills and technical skills. It is also one example of how social power relations (i.e. who is capable of managing accommodation profitability) have become more technologically dependent (see Jansson & Lindell, 2018: 1).

Taxation is another important topic provoking discussion among hosts. They want to be sure of how taxation works: what is considered an expense, what is considered as profit, whether there are any limits for tax responsibilities and what they are. Hosts also question whether Airbnb reports booking details to the relevant tax authorities. As taxation is one factor indicating the grey economy (e.g. the Finnish Tax Administration, 2018), those knowing and following the taxation rules are more likely to be part of society and the system, whereas those not paying taxes are under the risk of operating in the black market. This is ineluctably related to social division in societies. Even though challenges regarding unpaid taxes and the black market are not new phenomena and are not caused by the digitalised world or by the platform economy, the sharing economy platforms have brought a new dimension to the issue. In the case of Airbnb, by enabling individuals to participate in the accommodation industry, the platform shifts responsibility from companies and organisations to individuals, for example, in terms of tax paying. And, again, the ability to handle these issues in a socially advantageous way also has implications for the social life of an individual. Based on this, Skeggs' (2004) argument about today's societal class inequalities being reshaped and reproduced through the *privileges* of individualisation, could be complemented with the *obligations* of individualisation.

In general, Airbnb community discussions address many similar dilemmas that hosts face while refashioning themselves as entrepreneurs, thus indicating that running Airbnb accommodation is largely a professional business for many hosts. For example, one host asked for a professional Airbnb photographer to take pictures of his listing. There are also discussions about how arduous and demanding it is to be an Airbnb host. A Slovenian host, for example, described how tired she is of maintaining short-term rentals: she either wants to rent only on long-term rentals or to find a decent agent to look after her property on a daily basis. Many hosts advised her to set minimum stays and one, interestingly, suggested she find another superhost nearby for co-hosting and thus sharing responsibilities. This is again an example of networking skills among hosts, especially among superhosts.

While networking and operation management skills can be said to enforce social capital accumulation, along with the platform logic and architecture that conditions it (see Rasmussen, 2014), emotional capital is reflected in the discussions in a form of *discretion*. Hosts discuss, for example, the level of hospitality their guests want to receive. A Slovenian host described a situation in which his guest interpreted a warm welcome and hospitality as sexual harassment. Hosts also ask for advice for situations in which they are unable to welcome everybody but still want to maintain a polite and hospitable image – for example, how to inform guests about inaccessible apartments clearly enough without sounding unwelcoming. One host offered to check if pictures taken by other hosts look 'cool' enough. We also found an interesting discussion where the host describes a polite guest who, unfortunately, was smelling terrible: the host had been trying to clean the room for the next guest, but the room continued to smell. The host asked for advice in the community about writing a review for this guest that is polite but also tells other hosts truthfully about them. She received a lot of suggestions and examples for such a review, but also a comment reprimanding her: '… If you did not have the courage to talk to him about the smell while he was living with you, it is unfair to shame him publicly now'.

Discretion, in terms of considerate interaction with a guest (i.e. behaving in such a way as to avoid causing offense, embarrassment or excessive attention), also concerns the principles of hospitality and refers to one's hosting skills, in respect of which many distinctions are drawn among hosts. The following comment, made by a superhost, is an illustrative example of the ways of differentiating oneself (as a superhost) in terms of hosting.

> … it's the symbolism that is important! By offering a cheese and wine plate I am offering friendship to people. You and I, we have different ways of hosting, we are possibly looking for different things out of our hosting. I won't say I couldn't care about the money, it's nice but, we don't need it! To me, the hosting is a part of my life, not an impediment, or a necessity to it!

This quotation, as well as all the discourses described above, indicate the need for and the role of emotional capital in the performances of hosts. In acts of hospitality, discretion and the symbolism of small gestures that make guests feel welcome are manifestations of emotional capital – and they play a prominent differentiating role among platform users as hosts, gaining them an advantage in the hospitality stratification order.

Superhosts seems to be very active in the Airbnb community. They discuss engagingly about all themes, as well as about gaining and keeping their superhost status. The activeness of superhosts in the community probably indicates a generally high involvement and activity of superhosts in Airbnb itself. This is understandable because, to gain and to keep superhost status, they need to be proactive on many levels. However, concern about guest reviews (one of the criteria for superhost status) seems to be most evident in the community discussions, for example how to obtain better reviews and what the review variables actually mean from the host's point of view. In one illustrative example of the pressure of satisfying guests, a Finnish superhost relayed a story about an unsatisfied customer who was expecting 100% silk bedlinens, but did not get them. The host was worried about negative reviews and her reputation as a superhost.

In a way, the desire to gain and keep superhost status merges other themes we found in our analysis. As mentioned above, superhosts are encouraged to network in many different ways in the discussions. General management skills are primarily the skills that are rewarded through the superhost programme, and discretion is especially related to high-quality guest reviews. It requires both social and emotional capital to gain and to keep superhost status. In turn, superhost status supports the appearance of the host's listings when a user is searching for accommodation in a specific city. In this manner, the platform has the definitional power to define the desired host's traits and then accordingly reward them by its algorithms. Altogether, our observations stress the need for a socially sensitive understanding of tactics of how to navigate the digital algorithmic world of the platform economy and how to obtain and constantly cultivate the required types of skills and resources needed to prosper in such a business environment. As clear references to social and emotional capital appear in the online discussions, as illustrated above, the role of aesthetic capital is analysed in greater depth in the following section through an analysis of Airbnb host profiles.

Performing emotional and aesthetic labour: Analysis of online self-presentation

As noted by Roelofsen and Minca (2018), Airbnb suggests that the profile should be a space where providers describe, among other things, their daily routines and the intimate self that the host is willing to offer to the platform. They further argue that 'the production of the Self through such profiling practices shows how digital bodies are supposed to represent "real" and truthful bodies' (Roelofsen & Minca, 2018: 175). Following on from this, we are predominantly concerned with the ways in which a provider's identity is enacted via the profile's self-presentation by processes of symbolic differentiation because a renewed class analysis places much greater emphasis on processes of taste, lifestyle and culture (see Bottero, 2004) and thus on cultural differentiation. In their account of social relations, West and Fenstermaker (1995: 9) advance a reconceptualisation of 'difference' as an ongoing interactional accomplishment, performatively achieved through everyday interactions. In line with this, we considered online providers' profiles in terms of the interactions between providers and guests (not as merely textual pieces), as there is an obvious co-presence of both platform users involved in the performances of public self-presentations (see Picard & Buchberger, 2013).

The analysis of verbal content focused primarily on the level of professionalism indicated by the hosts and exploratory identification of the signs of emotional labour in their self-presentations, such as the ways in which and the extent to which the providers describe their personal life stories, extending from performing a professional persona to performing an image of a friend with whom the guests could form real connections, thus inviting guests into their intimate spheres.

Our analysis of visual material followed Manovich's Instagram-based analysis of competitive photography in social media (Manovich, 2017). Specifically, we focused on the identification of two embodiments of aesthetic labour and, concomitantly, signs of differentiation that indicate distinct user competences when it comes to representing the home through photo material: (1) identification of recurring visual conventions that are distinctive and (2) identification of distinguishable codes of staging. Distinctive visual conventions include, for instance: taking photos from an angled perspective or composition around diagonal (not vertical/horizontal) lines to make the ordinary seem extraordinary; clear differentiation between the main subject and the background; increased contrast; the use of point-of-view (i.e. looking down from above); detailed areas juxtaposed with empty areas in pictures, among other strategies. The distinguishable codes of staging refer, by contrast, to richness in detail; careful arrangements and aesthetic transformation of everyday life (showing expert judgement in matters of taste, ranging from pillows to textiles and table arrangements); staging appreciation for simple things by photographing a cup of tea or a book lying on the table in one's home; highlighting the professionalism of essentials, such as showing a high-quality coffee machine; or general avoidance of any stereotypical subject or representational style that is otherwise popular in commercial real-estate discourse.

We analysed the hosts who managed to be the most successful in Airbnb's search engine algorithm, as we assumed that these would be the hosts with the highest levels of experience and thus also high levels of emotional and aesthetic labour in their performance. This explains why the analysis of the accommodation providers' listings and profiles showed a very high level of professionalism in general. We concluded this from the facts that (1) all listings were entire spaces (no shared rooms), thus reflecting the position that Airbnb is a capital-based platform, (2) almost all accommodation hosts provide more than one listing, (3) all had an extremely high number of guest reviews (more than 100, in some cases even more than 1000), (4) almost all were so-called 'superhosts' and (5) all spoke at least one foreign language and, in some cases, were able to communicate in up to four or five languages.

In contrast to accommodation providers, the experience providers show a lower level of professionalism in terms of typical Airbnb indicators, such as the number of experiences listed (i.e. usually they offer only one experience). The experience providers generally have far fewer reviews than accommodation providers, reflecting the relative newness of Airbnb experience provision in these cities and thus less competition among the providers in terms of the difficulty of gaining a 'front page' position in Airbnb's search engine. The experience providers present their level of professionalism in a different way to the accommodation providers: they try to build their credibility via their descriptions. It is quite common for them to state they are providing experiences professionally, on other words, they make a living as tour guides or as tourism professionals and/or have their own company providing such experiences. Additionally, a common way to build credibility is to refer to education and professional knowledge in the area of experience (e.g. stressing that one has a PhD in the relevant topic or pointing out the level of education and/or the institution where they studied or worked).

The analysis of the personal profiles of hosts – both accommodation and experience hosts – showed an unexpectedly low level of elaborated personal details or personal stories. In the Finnish case, in particular, the personal descriptions within the profiles usually included only names, personal photography and sometimes extremely short personal descriptions. This is in contrast with Turnšek and Ladkin's (2017) predictions on extending the emotional labour of hosts by providing an intimate self versus the professional persona of a tourism employee. Although this was only an explorative analysis with the aim of providing insight into the emotional and aesthetic labour of Airbnb hosts, we did observe initial indications of cultural differences in terms of Finnish hosts being more reserved in their personal descriptions than Slovenian hosts. Rather than providing credibility via personal life stories, the hosts – especially the Finnish hosts, in line with their professionalism – more often accentuated their knowledge and expertise. That said, there are also instances of extremely elaborate and in-depth descriptions of hosts' life stories, accentuating their passion and hobbies and almost always pointing out their love for travelling. A common thread, identified especially within the descriptions of experiences, was that quite often the providers stressed the importance of changing one's life (e.g. career path) in order to be more true to oneself or to get in touch with nature – and thus providing the experience – together with a common emphasis on mindfulness and integration of its principles into the experiences.

Emotional and aesthetic labour was thus not so much identified within the personal profiles of the providers, even though their profile pictures were quite often of a personal nature (e.g. with family or pets) and were always made in such a way as to accentuate the physical attractiveness of the provider (one of the backbones of aesthetic labour). The experience providers accentuated the aspects of emotional labour in their descriptions of the experiences: a common way to describe an experience that they provide, for example, is to emphasise that the provider treats the guests as friends and promises that true connections will be built within the experience – in one case even proposing to stay in contact after the experience, alluding to building true friendships rather than the professional relationship of service provider and guest. Resourcing the self in such interactions implies a good measure of reflexivity, a resource rarely acknowledged as class-based. However, as argued by Skeggs (2004: 134), reflexivity and the techniques necessary for self-formation and self-telling are not equally available to everyone. The projects of the self are related to one's competence or cultural capital, which conditions the possibilities of reflexive individualisation. Reflexive projects of the self are thus always class formations that are dependent on the individual's possession of capital. The ability to effectively construct, represent and make use of the self through the hospitality platform therefore, on the one hand, reflects the existing power relations in society and, on the other, reinforces them.

Our analysis of visual content, following Manovich (2017), showed extreme diversity among providers in terms of stylisation of visual material. Given the nature of the service, accommodation providers have to strive much harder than experience providers in order to build on competitive differences with photographic material, since experiences offer a much larger possibility of diverse photographic material than accommodation. While a large extent of accommodation photography was highly stylised, we cannot conclude that this would be the norm for all photographic material. The extremely stylised photos, however, always accompanied the description of an apartment described as 'stylish' or similar – thus indicating that, in these cases, the aesthetics of the accommodation and its photos are presented as a whole. In some cases, photos are verified by Airbnb and in those instances every picture also has a short extra description. The Airbnb-verified photos looked very similar in terms of lighting, with bright, sunny light making the accommodation seem larger and more inviting. The experience providers use much more diverse photographic material than the accommodation providers. However, the aesthetic and visualisation skills seem to be taken as one of the self-understood skills, since a large number of experiences either build on photography (e.g. photography workshops, photoshoots or guided tours specially designed for photography) or the providers offer, alongside the main service, the additional service of taking photographs of guests. In line with this, the experience providers who include photography in their services stand out in terms of the techniques and skills used for distinctive visual representations of their services.

To conclude, Airbnb profiles are sites where the self is continually produced, represented and capitalised on, thereby constituting and reproducing differences and inequalities. Such a perspective exposes how platform architecture makes cultural differences among providers more perceptible and casts them into constant scrutiny, in this way contributing to the formation of an all-encompassing neoliberal culture in which the value of one's self is continually contested and one's culture is commodified.

Conclusion: Increasingly Open, Relentlessly Unequal

The inequality of the platform economy manifests itself in the traditional, familiar forms of unequal resources, such as the capital needed to provide more than one listing within the accommodations or high levels of educational achievements within the experiences. However, as we have shown, problematising platform work today from the perspective of inclusion should refer not only to the issue of access, but also to the issue of the unequal distribution of competences required to participate in the platform successfully. A more nuanced conceptualisation of capital dynamics should thus be taken into account when addressing the issue of inclusion in the platform economy. This does not mean that we underestimate material resources in any way. Rather, we maintain that in addition to economic capital, which is undoubtedly primary in the platform economy (distinguishing among those who have available material resources to share and capitalise on and those who do not), cultural capital and its subforms (emotional and aesthetic in particular) should also be taken into account as they condition proficient performance and navigation in the online hospitality business. Although differences regarding cultural resources – in particular, managerial, emotional and aesthetic skills – are not being seen as a true barrier to participation in the platform economy in general, the varying degrees should nevertheless be considered. They organise differences in providers' performances regarding their authenticity, responsiveness, discretion, care, flexibility and aesthetic manipulation in digitally representing their home (both visually and textually) in a market-appealing manner, as well as their operation management and networking skills. These cultural resources have potential to be accumulated over time, allowing those providers with higher levels of required cultural resources to form profitable distinctions and retain advantages in the networked hospitality business, in this way creating hierarchies and reinforcing social divisions. Platforms legitimate these resources as desirable dispositions, yet access to these resources is rendered structurally unequal. Platforms establish different statuses of providers and in this way form a distinct stratification order in the platform community.

Although necessarily limited, the discussion provided here is intended to prompt a consideration of the complex ways in which social space and technology interact. We stress that, in today's platform economy – characterised by inclusiveness and openness – battles against inequalities are not only battles over the material, but they are also battles over immaterial skills and competences, which are not being seen for what they are: socially produced distinctions. These are a product of differences in individual life opportunities and not a product of one's individualised hard work and natural talent. Under the idea of entrepreneurial self-fashioning, grounded in the myth of meritocracy, it appears that only successful (super)hosts naturally have what it takes to prosper in hospitality platforms. In this way, by naturalising socially produced distinctions, the platform economy helps to conceal and reproduce durable social inequalities. Specifically, the platform economy appears, at face value, to be more and more open, yet it in fact remains relentlessly unequal.

The future of inclusive tourism via the platform economy thus seems to continue to build on the traditional inequality of material resources, yet it is also faced with the distinctive issues of the unequal distribution of skills of emotional and aesthetic labour, combined with high levels of skills in visual communication and digital navigation of the world of algorithms.

References

Airbnb (2019a) How to start hosting. See https://www.airbnb.com/b/setup (accessed March 2020).
Airbnb (2019b) What is a superhost? See https://www.airbnb.com/help/article/828/what-is-a-superhost (accessed January 2021).
Baum, T. (2007) Human resources in tourism: still waiting for a change. *Tourism Management* 28, 1383–1399.
Bialski, P. (2012) Technologies of hospitality: How planned encounters develop between strangers. *Hospitality and Society* 1 (3), 245–260, doi:10.1386/hosp.1.3.245_1.
Bourdieu, P. (1984/2010). *Distinction: A Social Critique of the Judgement of Taste.* London and New York: Routledge.
Bottero, W. (2004) Class identities and the identity of class. *Sociology* 38 (5), 985–1003.
de Certeau, M. (1984) *The Practice of Everyday Life.* Berkley: University of California Press.
De Groen, P.W., Maselli, I. and Fabo, B. (2016) The digital market for local services: A one-night stand for workers? An example from the on-demand economy. Brussels: Centre for European Policy Studies.
Eyal, E., Fleischer, A. and Magen, N. (2016) Trust and reputation in the sharing economy: The role of personal photos in Airbnb. *Tourism Management* 55, 62–73.
Fagerstrøm, A., Pawar, S., Sigurdsson, V., Foxall, G. and Yani-de-Soriano, M. (2017) That personal profile image might jeopardize your rental opportunity! On the relative impact of the seller's facial expressions upon buying behavior on Airbnb. *Computers in Human Behavior* 72, 123–131.
Finnish Tax Administration (2018) Estimates on the shadow economy and tax gap. See https://www.vero.fi/en/grey-economy-crime/scope/estimates-on-the-shadow-economy-and-tax-gap/ (accessed March 2020).
Fitzmaurice, C.J., Ladegaard, I., Attwood-Charles, W., Cansoy, M., Carfagna, L.B., Schor, J.B. and Wengronowitz, R. (2018) Domesticating the market: moral exchange and the sharing economy. *Socio-Economic Review* 18 (1), 81–102, doi:10.1093/ser/mwy003.
Germann Molz, J. (2013) Social networking technologies and the moral economy of alternative tourism: The case of Couchsurfing.org. *Annals of Tourism Research* 43, 210–230.
Haanpää, M., Hakkarainen, M. and Harju-Myllyaho, A. (2018) Katsaus yhteiskunnalliseen yrittäjyyteen matkailussa: Osallisuuden mahdollisuudet pohjoisen urbaaneissa paikallisyhteisöissä (Review of social entrepreneurship in tourism: Possibilities of participation in the Northern urban communities). *The Finnish Journal of Tourism Research* 14 (2), 44–58.
Hakkarainen, M. and Jutila, S. (2017) Jakamistalous matkailussa (Sharing economy in tourism). In J. Edelheim and H. Ilola (eds) *Matkailututkimuksen avainkäsitteet* (pp. 183–187). Rovaniemi: Lapland University Press.
Hochschild, A.R. (1983) *The Managed Heart: Commercialization of Human Feeling.* Berkeley, CA: University of California Press.
Igarashi, H. and Saito, H. (2014) Cosmopolitanism as cultural capital: Exploring the intersection of globalization, education and stratification. *Cultural Sociology* 8 (3), 222–239.
Jansson, A. (2016) Mobile elites: Understanding the ambiguous lifeworlds of sojourners, dwellers and homecomers. *European Journal of Cultural Studies* 19 (5), 421–434.
Jansson, A. and Lindell, J. (2018) Media studies for a mediatized world. rethinking media and social space. *Media and Communication* 6 (2), 1–4.
JPMorgan Chase Institute (2016) *Paychecks, Paydays, and the Online Platform Economy: Big Data on Income Volatility.* See https://www.jpmorganchase.com/corporate/institute/document/jpmc-institute-volatility-2-report.pdf (accessed March 2020).

Khan, S.R. (2011) *Privilege: The Making of an Adolescent Elite at St. Paul's School*. Princeton, NJ: Princeton University Press.

Lee, M.K., Kusbit, D., Metsky, E. and Dabbish, L. (2015) Working with machines: The impact of algorithmic and data-driven management on human workers. In B. Begole (ed.) *Proceedings of the 33rd Annual ACM Conference on Human Factors in Computing Systems* (pp. 1603–1612). New York, NY: Association for Computing Machinery.

Littler, J. (2018) *Against Meritocracy: Culture, Power and Myths of Mobility*. Abingdon: Routledge.

Liu, S. and Mattila, A. (2017) Airbnb: Online targeted advertising, sense of power, and consumer decisions. *Journal of Hospitality Management* 60, 33–41, doi:10.1016/j.ijhm.2016.09.012.

Manovich, L. (2017) Instagram and contemporary image. See http://manovich.net/index.php/projects/instagram-and-contemporary-image (accessed January 2021).

Mijs, J.J.B. (2018) Earning rent with your talent: Modern-day inequality rests on the power to define, transfer and institutionalize talent. See https://osf.io/preprints/socarxiv/ky9vh/ (accessed January 2021).

Picard, D. and Buchberger, S. (2013) *Couchsurfing Cosmopolitanisms: Can Tourism Make a Better World?* Bielefeld: Transcript Verlag.

Rasmussen, T. (2014) *Personal Media and Everyday Life*. New York, NY: Palgrave Macmillan.

Roelofsen, M. and Minca, C. (2018) The Superhost: Biopolitics, home and community in the Airbnb dream-world global hospitality. *Geoforum* 91, 170–181.

Savage, M., Warde, A. and Devine, F. (2005) Capitals, assets, and resources: some critical issues. *The British Journal of Sociology* 56 (1), 31–47.

Schmid-Drüner, M. (2016) *The Situation of Workers in the Collaborative Economy*. Brussels: European Parliament.

Schor, J.B. (2017) Does the sharing economy increase inequality within the eighty percent? *Cambridge Journal of Regions, Economy and Society* 10 (2), 263–279.

Skeggs, B. (2004) *Class, Self, Culture*. London: Routledge.

Striphas, T. (2015) Algorithmic culture. *European Journal of Cultural Studies* 18 (4–5), 395–412, doi:10.1177/1367549415577392.

Teubner, T., Saade, N., Hawlitschek, F. and Weinhardt, C. (2016) It's only pixels, badges, and stars: On the economic value of reputation on Airbnb. In *Proceedings of the 27th Australasian Conference on Information Systems (ACIS2016)*. University of Wollongong Faculty of Business. See https://ro.uow.edu.au/acis2016/papers/1/56/ (accessed March 2020).

Turnšek, M. and Ladkin, A. (2017) Changing employment in the sharing economy: The case of Airbnb. *The Public* 24 (sup1), S82–S99.

Veijola, S., Hakkarainen, M. and Nousiainen, J. (2013) Matkailu työnä (Tourism as work). In S. Veijola (ed.) *Matkailututkimuksen lukukirja* (pp. 173–185). Rovaniemi: Lapland University Press.

Warhurst, C., Nickson, D., Witz, A. and Cullen, A.M. (2000) Aesthetic labour in interactive service work: Some case study evidence from the 'new' Glasgow. *The Service Industries Journal* 20 (3), 1–18.

West, C. and Fenstermaker, S. (1995) Doing difference. *Gender & Society* 9 (1), 8–37.

Zervas, G., Proserpio, D. and Byers, J. (2015) A first look at online reputation on Airbnb, where every stay is above average. See https://papers.ssrn.com/sol3/papers.cfm?abstract_id=2554500 (accessed March 2020).

Part 2
Methods

3 Inclusion in Tourism Strategies: Setting the Stage for Inclusive Tourism Development in Tourism Destinations

Anu Harju-Myllyaho and Salla Jutila

Introduction

Inclusion, as a term, relates to something that integrates us to become part of society, regardless of who we are. Biddulph and Scheyvens (2018: 584) define inclusive tourism as transformative tourism engaging marginalised groups in the production and consumption of tourism and the sharing of its benefits. Thus, inclusion stresses both guest and host viewpoints. Accessibility, in turn, emphasises the consumer's point of view. According to the United Nations World Tourism Organization (UNWTO, 2016), accessible tourism requires any tourism product to be designed irrespective of age, gender or ability, and with no additional costs for customers with disabilities or specific access requirements. Thus, accessible tourism can be seen as part of inclusive tourism. The aim of this chapter is to analyse how inclusive tourism futures are constructed in tourism strategies.

According to Dredge *et al.* (2011: 3), in Western economies, tourism planning and policymaking are linked to changes in the ideological and sociopolitical landscape. Western governments' commitment to neoliberal economic management and globalisation has focused on increasing growth in the name of economic well-being. Moreover, Dredge *et al.* (2011: 2) state that, as social processes, tourism planning and policymaking can be affected by personal and/or group values, interests and ideology. The approach in this chapter is based on this line of thought, which is then continued with notions of the policy metaphors developed by Vallis and Inayatullah (2016). According to them, the ways we perceive things and think and act are – in the end – metaphorical. The way we experience the world is thus metaphorical. However, like so many things that take place in our daily lives, the metaphors become invisible (Vallis & Inayatullah, 2016: 1). In terms of policymaking, it is important to search for deeper insights into the processes by seeking understanding of the underlying metaphors. Indeed, according to Vallis and Inayatullah (2016: 2), powerful metaphors not only describe our reality, but also build it. Consequently, this would mean that in order to build inclusive tourism futures, close attention should be paid to the way they are articulated.

The research material used for this chapter was tourism strategies and policy papers from areas that can be regarded as peripheral. The areas under examination are Finland, Lapland, Portugal, Madeira, Scotland and the Scottish Highlands and Islands. These regions were chosen because they represent popular tourism destinations located far from the European centre.

The strategies were analysed using causal layered analysis (CLA) as a method. CLA is a futures research method that has also been used in strategic planning. It provides a tool for challenging conventional thinking, facilitating multidimensional conversations and gaining a better understanding of our own worldviews and sense-making processes. CLA can also be used in constructing shared visions and ensuring that existing plans include deep and diverse perspectives (see Inayatullah, 2004) In the work for this chapter, CLA was used for (1) recognising and deepening our understanding of the ways in which tourism strategies guide thinking and (2) identifying the underlying worldviews behind the strategies themselves and how they affect the way inclusion is seen in the tourism industry. CLA can be used in various ways, as pointed out by Vallis and Inayatullah (2016: 2). Our aim here is to study the way inclusion is perceived and articulated in tourism strategy papers. Alternative futures are not built within this chapter – instead, we leave the task of imagining different futures to the reader.

In general, the purpose of strategy papers is to pave the way to different futures and, as such, they can be considered as actions that construct the future. The futures images proposed in tourism strategies are projections of the dominant values in the present, and those values are realised through conventions and actions. Kamppinen and Malaska (2003: 98) write that the future has a different ontological status than the past and the present. The future is neither here for us to observe, nor does it exist as memories; consequently, its ontological status is mental (Kamppinen & Malaska, 1999: 98). Thus, strategy papers, which aim to prepare for the future, are based on the mental futures images of their creators. Futures images, or images of the futures, are images of the various possible futures (Sitra, 2019). De Jouvenel (2019) uses the term 'futuribles' to describe these alternative images, which are partially known (or at least somewhat certain) and in part imaginary (see De Jouvenel, 2019; Söderlund & Kuusi, 2003: 281).

Inclusion and Hospitality

In tourism, various terms and concepts relate to inclusion, such as inclusive tourism, tourism for all, universal tourism, barrier-free tourism and accessible tourism (Darcy & Buhalis, 2011a: 10). All these concepts have the same goal, which is to transform tourism from an exclusive luxury that can be only afforded by a few individuals to an inclusive activity that is open to all. Affordability in this case is more than an economic concept – it also includes social and cultural capital, for instance. Accessible tourism is the most common term that is used when referring to inclusiveness from a tourist's point of view. One aspect of accessibility is surely economic. However, other barriers might also prevent people from taking part in tourism. In addition to economic barriers, there are also physical, political, social or cultural barriers, as well as communicative barriers. These are all interrelated. For instance, a person might have a disability, which can also lead to economic downfall, preventing tourism and leisure activities. Awareness and reverence of difference, as well as the desire to serve every tourist, can be seen as aspects of socially and culturally accessible tourism, that can significantly remove physical barriers or barriers in communication (Darcy & Buhalis, 2011b: 27; Jutila & Harju-Myllyaho, 2017: 225). Harju-Myllyaho and Kyyrä (2013: 16) use the term accessible hospitality to describe their attitude and approach.

Sedgley *et al.* (2012) write about tourism poverty, which in the contexts of this chapter can be understood both from the viewpoint of the tourist and from the viewpoint of the host. A person who is in one way or another restricted from making decisions might not be able to start a company or even work in the tourism industry. The reasons for this can vary in a similar way to if the person was a tourist coming to a destination. There can be economic, social, cultural or even political factors that limit freedom of choice. Sedgley *et al.* (2012) write that tourism poverty is often discussed in the context of developing countries, but is not commonly addressed in more affluent countries.

Although inclusion is not a synonym for freedom, from the viewpoint of freedom of choice, the two terms are very much related. Tourism as an activity has become quite a common part of life, at least in Western societies. This means that being excluded from tourism is one aspect of being excluded from daily life (Harju-Myllyaho, 2018: 23–24).

Even though there are innumerable amounts of people who for one reason or another are not able to travel, tourism can be seen at least as a social right, if not a human right (McCabe *et al.*, 2012: 3). This viewpoint is based on the Universal Declaration of Human Rights, which states that all people have the right to movement and leisure (United Nations, 1948). It has to be pointed out, though, that the declaration was proclaimed in 1948 and the times then were quite different compared to today.

Post-structuralist Approach to Inclusion

Inclusion is an umbrella term with various viewpoints. These viewpoints make specific definitions challenging, but they also help to form multifaceted and multidisciplinary conversations (Isola *et al.*, 2017: 3). According to Isola *et al.*, inclusion consists of involvement, relatedness, belonging and togetherness. It also relates to participation, representation and democracy. In project Sokra (a coordination project for social inclusion), inclusion is seen as being part of an entity, where a person can attend to different sources of wellness and create meaningful relationships. Thus, inclusion is being able to influence one's own life, possibilities, activities services and certain decisions (2017: 3). In tourism, inclusion often refers to accessible tourism – in other words, the possibility to make the decision to travel without limitations. Tourism creates possibilities, but they are not equally available to all. Inclusion means not only the possibility to travel and participate in tourism as a consumer, but also the possibility for locals to take part in decision-making and tourism development.

The way inclusion is seen in the context of this chapter is based on the idea that inclusion is partly formed around a sense of welcoming (e.g. Höckert, 2015), which we all perceive in different ways depending on who we are, our identity or our subject position. Höckert (2015) used hospitality as a way of approaching participation in tourism. She writes that both terms – hospitality and participation – require openness towards others (2015: 91). According to Selwyn (2000: 19), besides the obvious (serving food, drinks and accommodation), hospitality includes the notion that its function is to promote an established relationship. Therefore, hospitality acts symbolically affirm structures of relations or transform such structures. Selwyn (2000: 19) states that hospitality converts strangers into acquaintances, enemies into friends and friends into even better friends, and points out that these principles apply in ethnographic descriptions of a wide variety of social systems.

According to Isola *et al.* (2017: 3) goal-oriented agency and a 'wilful mind' are born in the interaction of inner and outer possibilities. Inner possibilities include, for example, salaries, social security, living conditions, education and hobbies. Outer possibilities include received education and learning, as well as individual talents and personal history. Inclusion has also faced some criticism. According to research, inclusion as participation, in particular, has been under criticism, since it has been seen as 'forced participation' (Isola *et al.*, 2017: 3).

Post-structuralism as a theoretical approach complements structuralism by adding a viewpoint that argues that, even though we acknowledge that our view of the world is socially structured, the way we see and study those constructions is also constructed in a certain way (Inayatullah, 1990: 129–130). Inayatullah (1990) sees the post-structuralist approach as leading us to consider critically the trends that are given to us by traditional futures research literature: 'The issue is not only what are other events/trends that could have been put forth, but how an issue has been constructed as an event or trend in the first place …' (Inayatullah, 2004: 12). Inayatullah also notes that taking different epistemological viewpoints, such as those of different civilizations, is one of the most useful ways of creating distance from the present (Inayatullah, 2004: 12). The post-structuralist viewpoint suggests that even if we write about inclusion and acknowledge it, it is important to note that the way we perceive inclusion is also already structured and, in a way, restricted since it is based on certain presumptions. We thus come to the notion that we should in fact examine the epistemological assumptions of the real when discussing planning efforts such as strategy papers.

Post-structuralist thought problematises the existing *status quo* and conventional thinking by highlighting socially and culturally constructed viewpoints and mental models that are inherited and built within us in a profound way. In other words, inclusion and exclusion gain ground in the socially constructed way of seeing the world, understanding what is normal and leaving out things that do not fit to the frame of normality. Social and cultural norms define who is accepted as a full member of society. Following Isola *et al.* (2017), inclusion is built on social background, economic affluence, gender, ethnic background and physical disabilities, for instance.

Tourism Strategies and Inclusion

In a Delphi study, Harju-Myllyaho and Jutila (2018) noted that developing accessible tourism (i.e. customer inclusion) is dependent on how society, community and industry understand difference and are committed to defending equality and equity. Furthermore, it is a question of how people with access needs are taken into consideration in development work. Behind the decision-making system, there are different opinions, worldviews and ideologies, which guide our behaviour and attitudes towards others and the world in general. These are manifested in the ideologies of political parties, for instance. Exploring the 'meta level' deepens our understanding of people's personal and common beliefs that belong to them by nature. At this level of myths and metaphors, the views are so deeply rooted in history and culture that they are difficult to change, or they change only very slowly.

The tourism decision-making process is by no means rational. It is often thought that political decisions are justified with reason, that they are somehow rational. However, these important decisions are also made with ideologies, intuition and 'gut feeling' (Dredge *et al.*, 2011: 2; Hall, 1994: 3). Hence, if we wish to understand the problems and bottlenecks of tourism inclusion, we must first study the ideas behind them. It is noteworthy that tourism policymaking is based on both ideologies and intuition, and functions as a question of power. The political framework in which decision-making takes place should be acknowledged and, to do so, various theoretical viewpoints and academic traditions are required (Hall, 1994: 14).

As noted by Hakkarainen and Tuulentie (2008: 5), the strategy process always has its conventions and representations. According to these authors, tourism strategies not only describe certain systems, but are also actively creating them. In addition to paying attention to what is written in the strategies, we should also note what has not been said and what has been excluded. The power of these documents is important to consider (Hakkarainen & Tuulentie, 2008: 5–6).

Inayatullah (1990) writes that, in traditional strategic thinking, time is something that must be tamed. Unless we have knowledge about the future, we lose market shares, are unable to acquire funding or fail in fulfilling the requirements of a modern institution. Usually, the general aim of a futures project is to seek better quantitative, dynamic or otherwise understandable ways to form a picture of the future possibilities to be able to make better decisions. The methods must be scientific and objective, and the researcher should not influence the results. Planning is seen as an elemental part of modern Western society and as a way towards the desired world but, as Inayatullah (1990: 116) states, 'every effort to plan the future is submerged in an overarching politics of the real'. In other words, building strategies and planning activities belong to the liberal Western world, where progress is measured by numbers and where the future is tamed and steered with strategies and plans. Planning is a sign of a progressive society. However, the way our societies and thoughts are organised might hinder and challenge the planning – or at least realisation of the plans.

Globalisation and the economic order (constant demand for growth) create homogenised ways of thinking; consequently, this renders different viewpoints coming from other cultures less believable. As Inayatullah (1990: 137) writes: 'If we don't transform, we create the used future'. A critical approach reveals some of the structures of power and enables the deconstructing and reconstructing of possible pasts, presents and futures (Inayatullah, 1990: 137).

Research Material

The research material for this study was tourism strategies and policy papers from areas in Europe that can be seen as peripheral – Portugal/Madeira, Scotland/Scottish Highlands and Islands and Finland/Lapland. Portugal is located in the southwest corner of Europe. It also includes Madeira and the Azores, both located in the Atlantic Ocean. Madeira is a 1.5-hour plane flight from Lisbon (Visit Portugal, 2019). Scotland is situated in the mid-west of Europe and is the northernmost part of the United

Kingdom. It has some 800 islands, including Shetland and Orkney (STA, 2019). Finland, located in the northeast part of Europe, is a considerably large geographical area. The distance between the northernmost and southernmost points is 1157 km. The average population density in the country is 18.1/km^2 (Visit Finland, 2019). Geographically, Lapland covers almost one-third of the area of Finland. In terms of population, only slightly more than 3% inhabitants of Finland live in the northernmost region (Statistics Finland, 2019).

Despite the distances between these countries, there are common issues that concern them. One common factor is that these countries have remote areas and nature attractions. On the other hand, they are different in terms of history, culture and social structures. Thus, the analysed strategies all belong to areas that are characterised by accessibility challenges and unique, yet fragile, environments. They are also dependent on tourism at least in the sense that tourism is considered as a provider of work and livelihoods in sparsely populated areas. From social and ecological perspectives, the role of tourism in these regions is significant. Nevertheless, the purpose of this study is not to compare the strategies, but to understand the picture of the present that the strategies are painting in terms of inclusive tourism.

The Portuguese tourism strategy is compiled by Turismo de Portugal (TDP), which is a public institution under the Ministry of Economy. The responsible actor for Madeira's strategy is Secretaria Regional da Economia, Turismo e Cultura (SRETC), under the regional government of Madeira. Scotland's tourism strategy was established by the Scottish Tourism Alliance (STA), which is the representative body of the Scottish tourism industry. The tourism strategy of the Highlands and Islands is created by the Highlands and Islands Enterprise (HIE), an organisation integrating economic and community development with a remit from the Scottish government. The responsible official for the Finnish national tourism strategy is the Ministry of Economic Affairs and Employment and Lapland's tourism strategy is devised by the Regional Council of Lapland (RCL). These strategies are described in greater detail as part of our CLA in the litany level of our analysis.

Causal Layered Analysis

According to Inayatullah, CLA is a method for opening the past and present to create alternative futures. It enables the horizontal spatiality of futures and vertical layers of analysis (Inayatullah, 1998: 815). In this chapter, we use the verticality layers to explore the different ways of knowing about the futures (i.e. the epistemic and methodological approach), which are not predetermined and thus remain unknown (De Jouvenel, 2019: 15) (understanding the ontological status of the future). We see horizontal spatiality as a wider perspective on the possible futures.

The benefits of CLA include, for instance, the fact that when used in a workshop setting, it leads to the inclusion of different ways of knowing among participants and moves the discussion to a deeper level. CLA not only tries to make predictions based on calculations, it aims to problematise the units of analysis (Inayatullah, 1998: 816–817). This is a very important viewpoint considering that tourism is a phenomenon with various impacts besides economic ones. Thus, it is important to identify new ways to measure the impacts of tourism.

CLA is a method that facilitates the examination of fixed mental models and provokes new ideas. According to Inayatullah (1998: 816–817), the aim of critical research is to disturb existing power relations through problematising our existing categories and evoking other places or scenarios of the future. CLA draws from post-structuralism as it emphasises the deconstruction of texts and their meanings (Inayatullah, 1998: 818). In practice, CLA has previously been used in exploring Singapore's national tourism strategy, young people's images of the future (Kaboli & Tapio, 2017) and value co-creation (Ketonen-Oksi, 2018), to mention just a few recent studies.

As already noted, in this study, the texts to be deconstructed are strategy papers. Moving between the different layers of the texts (tourism strategies in this case) allows for a deeper understanding of the discourses that have been dominant constituting the present. Post-structuralism, and thus CLA, allows researchers to ask questions such as 'Which future is privileged? How has the present been constructed? What could be the genealogies of the future?' (Inayatullah, 1998: 818). All these questions can be asked in terms of tourism strategies and inclusion.

Based on the theoretical discussions above, the concepts and issues that we see as part of inclusive tourism development were gathered. These concepts form the framework for the CLA (Table 3.1). The tourism strategies are reflected against this framework on different levels of the analysis. For the analysis, the different viewpoints were simplified to key terms: host–guest relationship, accessibility, participation, inner and outer possibilities, and hospitality. Table 3.1 can be read horizontally or vertically; in this chapter, we use the vertical perspective.

Based on Table 3.1, the chosen tourism strategies are described in more detail in the following. In the litany level, the obvious contents of the strategies – those that can be seen without conducting a deeper analysis of the content, meanings or connections – are described. In the systemic level, the societal structures and systems of tourism in the strategies are discussed.

Litany

According to Inayatullah (1998: 820), 'litany is the first level, which consists of quantitative trends presented by news and media. This layer

Inclusion in Tourism Strategies   67

Table 3.1 Literature-based framework for CLA – analysis

Category	Layer			
	Litany	Systemic	Worldview	Metaphor/myth
General	• Common policy talk	• Policy-centred, policy-driven • Tourism is an industry	• Liberal, growth-based • The economy is based on growth, tourism needs growth, tourism should grow	• Leading the market • 'We weren't born to follow' • Winning the competition for customers
Host-guest relationship	• Tourism is developed to benefit the regions • Locals are a part of the strategy	• Tourism as a system is exclusive and people are invited or included in it • The system includes the tourism actors	• Western, customer-oriented and partly customer-centred • Focus on foreseeing the needs of the customer	• There is a line between guests and hosts; front desk and backstage • Both locals and tourists are invited to the strategy as guests, making the strategy one of the many stages where tourism takes place
Accessibility	• Accessibility is seen often as accessibility to the regions (flight routes etc.), though there are some exceptions	• Tourism as a system is dependent on accessible routes • Transportation is a part of the tourism system	• Tourism is seen as everyone's right	• 'Yes you can' • Everyone is able to travel
Participation	• Collaboration • Diverse contributions • Open and participative • Collaboration is encouraged to better serve customers	• Tourism is mainly detached from other industries • Connection to other industries is seen in terms of product development and land use • As part of the tourism system, locals are seen as targets of the system	• People are active and participating by nature. They have the capacity and capability to participate • The chances to participate in tourism are equal • Participatory approach is very limited in terms of who is invited to participate	• 'The customer is always right' • 'The customer is the most important participant'
Inner and outer possibilities; possibility to impact one's future	• The possibilities of tourism are seen through economic lenses	• Tourism relies on the current economic system/order • By developing tourism, regions and people benefit and have better prospects	• Western, liberal and individualistic: everyone has equal possibilities to take part in tourism	• People want to be employed in tourism. Tourism work is desirable. • 'Anyone can do it' • Tourism creates possibilities
Hospitality	• Hospitality is customer service	• Hospitality is a one-way road: hosts produce hospitality, customers consume it	• Caretaking, customer-centred • Hospitality is written in practices and conventions, and also in policy work	• Locals are unique in their hospitality • 'Special hospitality' • Ethos of hospitality and participation

does not consist of holistic or connected elements, but instead, they appear disconnected and discontinuous'. It is the 'official unquestioned view of the reality' (Inaytaullah, 2004: 8).

In this study, this level consists of the obvious content of the strategies, describing the strategy papers on a superficial level for the reader to understand the content and structure of the strategies. Thus, in this level, we have not sought to conduct a deep analysis of the strategies, but rather to describe the strategies as they are presented. In general, the litany level in the strategies follows a common strategy language or even a pattern. The timespans vary between the strategies, as they are written for different periods and in different years. However, all the strategies were valid when the analysis was conducted.

The Portuguese tourism strategy was written in 2017 and is valid until the year 2027 (TDP, 2017). It is a 66-page document in Portuguese. TDP also provides a summary (32 PowerPoint slides) in English (TDP, 2016). Due to language barriers, the English version was mainly used for analysis in this study. Madeira's tourism strategy is for the years 2017–2021, amounting to 136 pages in Portuguese (SRETC, 2016). None of the authors of this chapter speak Portuguese, but Google Translate and other available sources were used to understand the content.

Scotland's tourism strategy was published in 2012; it is 18 pages long and addresses the strategy until the year 2020 (STA, 2012). The HIE is an organisation with a remit from the Scottish government that integrates economic and community development. As HIE has defined tourism as one of the crucial, growing sectors in the area, they also have determined the aims and focus of the tourism industry over the course of the years 2016–2019. The HIE website (HIE, 2019) was considered as the official tourism strategy of the Highlands and Islands in this study. After the analysis process for this study had already been completed, the STA also produced a separate action plan for the Highlands and Islands (STA, 2019) for implementing the Scotland tourism strategy. However, this action plan supports the results of the analysis of the website.

The tourism strategy for Finland is 43 pages long, with all attachments, and is valid until 2020 (Visit Finland, 2010). The first version of the strategy was published in 2006, but it was updated in 2009 and 2010. The Lapland tourism strategy was valid for the years 2015–2018 (RCL, 2015); a new strategy was being updated during the period of our analysis (spring and summer 2019) and we therefore viewed the initial policy as still valid. Despite the short time period covered, Lapland's tourism strategy is 59 pages long. The tourism strategy for Madeira is longest in terms of pages, while Finland's strategy covers the longest timespan.

The policy structure is similar in each strategy. The papers start with *status quo* analysis, including strengths and challenges. The *status quo* provides a basis for setting objectives and goals that are followed by steps to achieve the objectives. Additionally, cross-cutting themes of priorities are defined in each strategy, as well as plans for measuring and monitoring the success of the strategy.

In the strategy papers, tourism as an industry is described and measured in numbers and volumes. For instance, the Highlands and Islands action plan for tourism (STA, 2019: 4) states the following.

> We will use a robust set of key performance indicators to measure this: Dedicated Highland research using the 'DREAM' model which includes:
> - Direct spend
> - Indirect spend
> - Visitor numbers
> - Tourism employment

In some of the strategies, inclusion is mentioned briefly in terms of accessibility or local participation. Accessibility refers mainly to possibilities to reach the country or region. None of the strategies has a clear focus on inclusion.

Sustainable tourism has become a part of tourism development and current trends are moving towards more inclusive tourism development in terms of local people. In addition, climate change issues act as drivers for ecological sustainability. According to Dredge *et al.* (2011: 4), sustainability as a viewpoint has been a part of tourism planning, along with the dominant neoliberal discourse. This is seen in the strategy papers as well. For instance, the tourism strategy action plan of the Scotland's tourism strategy (STA, 2012: 14) explicitly states the following.

> Sustainable tourism
> With sustainable economic growth as our goal, we must seek to maximise our operational efficiency and environmental performance, minimise our impact on the local environment and connect with our communities to deliver real benefits.

Sustainable tourism is seen as a significant part of the tourism industry and a basis for growth in Scotland's tourism strategy (STA, 2012: 15). In the Madeiran strategy, there is a focus on sustainable tourism and committing to the ethical values of the World Code of Tourism Ethics (SRETC, 2016: 15).

Systemic/social causes

Inayatullah (1998: 820) writes 'in this level, interpretation is given to quantitative data and the analysis is often provided by policy institutes…'. 'The data or the research material can be explained more deeply or questioned in this level' (Inayatullah, 2004: 8).

In the litany level, we began the discussion of how tourism is described in terms of numbers and volumes. In the systemic level, we discuss how the revenues, overnights and target groups – among other things – form a complex system we call tourism and how they describe inclusion as a part of this system.

The strategies describe the tourism system as an industry, which is measured by the number of participants (tourists and overnights) and revenue. On a systemic level, it is notable how the strategies describe tourism as an industry, which is concomitantly a part of the modern market economy and a closed system detached from other industries. Peculiarly, as a system, tourism seems to be policy-oriented or at times even policy-centred. The Scottish tourism strategy states:

> Building on the good work being done by local tourism groups, local authorities and others, this action plan will provide a fresh challenge to all of us in the industry, with the STA playing a central role in encouraging businesses, destinations, sectors and other stakeholders to get involved. (STA, 2012: 12)

It can be said that inclusion cannot be found in the strategies on a systemic level, even though it is mentioned occasionally, for example accessible hospitality in Lapland's strategy (RCL, 2015: 26). The strategies do not suggest systemic solutions for changing the system to be a more inclusive industry, nor do they describe tourism as an inclusive system. Lapland's tourism strategy also describes the development process as follows.

> Nine regional meetings were organized during the process in different parts of Lapland. Great number of tourism entrepreneurs and developers took part in these events and gave their input to the strategy. In addition, a number of stakeholder meetings were organized. (RCL, 2015: 4)

As a system, tourism tends to be exclusive in nature. This is seen, for instance, in the way that tourism destinations and companies include different target groups and simultaneously exclude others. This is natural and unavoidable for the growth of the businesses and the entire industry.

The Portuguese strategy mentions accessibility. It is proposed in the strategy that one of the goals should be to strengthen accessibility to Portugal and within the country (TDP, 2017: 9). Therefore, inclusion appears in the form of geographical accessibility. The understanding of the participation process, however, seems to be rather narrow: 'Participation process, expanded by diverse contributions, including players from the main markets' (TDP, 2017: 3).

On a systemic level, inclusion can be understood as related to participation, but those are seen as part of the system are already involved in tourism and have existing status as 'key players'. It is also thought that, with the support of public authorities and economic incentives, tourism can be developed towards the desired, more profitable direction. On the other hand, the strategies call for the active participation of local inhabitants (e.g. RCL, 2015: 54).

On a systemic level, it can be seen that the strategies 'speak' to the tourism stakeholders who are able to understand the language of the strategies, who can interpret overnights, revenues and target groups and who are aware of documents such as the World Code of Tourism Ethics and their contents. The entrepreneurs, developers and 'key players' are seen as the stakeholders who provide input to the strategy and to the industry. The output is the products provided to the main markets and, in the end, the growth of the businesses.

As discovered in the litany level, sustainability is one of the key issues in the strategies. However, even if sustainability is a building block in the strategies, it is unclear how it is followed through and how it is measured. In the case of the Scotland tourism strategy, sustainable tourism is considered a significant part of the industry, but then again, in the action plan that followed (for the Highlands and Islands), it is mentioned as an issue to be developed, but there are no measures or means described in the document.

Worldview

According to Inayatullah (1998: 820) 'the third level takes the analysis to a deeper level, since it concerns the structure and the worldview that legitimates or supports the structure. On this level, one can study different discourses (economic, cultural, religious etc.) and how they actually do more than cause or mediate the issue. They constitute it. By drawing from different discourses, various scenarios can be created'.

The third level therefore consists of the unconscious, ideological assumptions and explores how different stakeholders see and construct the litany (Inayatullah, 2004: 8). In this study, we used this level to understand the cultural standpoint of the strategies and the dominant ideologies beneath the obvious content of the strategies. It has to be kept in mind that our realities are constructed through our own mental images. According to Inayatullah (2004: 11), the future is patterned and chaotic and it can be known and unknown at the same time. This means that even if the future is unknown – that even if everyone has the possibility to impact the future – the structures define the individual and, thus, create restrictions that prevent the individual from making a choice (Inayatullah, 2004: 11).

In the tourism strategies, the worldviews are dominantly liberal and growth-based. Even if defined as a mature tourism destination, the objectives in the strategies are to strengthen the growth cycle of the destination, as in the Madeira tourism strategy (SRETC, 2016). It appears in the strategies that a successful tourism industry recognises the role of locals and hosts. However, the strategies focus on customers and satisfying their needs. This, in the end, however, brings prosperity to locals, as well.

> A strong tourism sector in the Highlands and Islands can help create more resilient rural communities. If we get it right, tourism will help support a vibrant regional identity which attracts people to live, work, invest and visit our region. (HIE, 2019)

A liberal standpoint not only seeks growth in numbers, but also assumes that hosts and locals are able to engage with tourism activities and that they wish to do so. A liberal management has been, in the latter half of the 20th century, a part of Western governmental politics, as pointed out by Dredge and Jenkins (2011). They add that sustainable tourism has become a part of the discussion, but that it has remained on a rhetorical level rather than becoming a set of guiding principles or actions (Dredge & Jenkins, 2011: 3–4).

Even though the strategies are mostly participatory in nature, locals and hosts are seen as invited guests to the tourism scene (as described in Höckert's (2015) dissertation). For instance, the Scottish strategy clearly states that tourism development must be taken into consideration by tourism actors *and* also by the companies who do not seek growth or expansion (STA, 2012: 13). This means that the ethos of growth touches everyone – including those who do not consider themselves to need it. Consequently, having ownership of the development process does not belong to the participating local actors. Instead, they are invited to support the cause even if they are not particularly interested. The Finnish strategy suggests moving school holidays in order to improve conditions for tourism companies – schoolchildren and their families are not speaking out about what this would mean for them.

> Many companies offering summer tourism services close their doors in the middle of European holiday season. This decreases the attractiveness of Finland as a summer destination, profitability of tourism companies as well as tax revenue. Moving summer holidays forward with two weeks would significantly improve the situation. (Visit Finland, 2010: 31)

Participatory development is a considered aspect in each of the strategies. However, it seems that the time of the participation takes place in the implementation phase. In other words, some stakeholders are invited to implement the strategy, but their part has already been written for them. As the Madeiran strategy states, stakeholders 'feel involved and identify with the strategy' (SRETC, 2016: 64). Thus, everyone is not involved in the strategy process, even though they are considered to be an important stakeholder group or even a profound part of the industry's offerings. Indeed, in Lapland and Scotland, for instance, the emphasis is on the exceptional hospitality of the local people (RCL, 2015: 26; STA, 2012: 5).

It is also worth noting that 'relevant stakeholders' is an expression used in each strategy, but none of the strategies reveal how and why certain stakeholders are actually seen as relevant while others are not.

Relevant stakeholders include those considered to be 'important', while others are excluded. The stakeholders mentioned most frequently in the strategies are distinctly different authorities and officials. Regional organisations, destination management organisations and producers of statistics are examples of the most emphasised actors. Tourism companies, research institutes, associations and municipalities, as well as tourism resorts, are also present in each strategy, but they receive less attention than the authorities. Local people, as well as other livelihoods (besides tourism), are mentioned only a few times in the strategies, but they are positioned as objects of the discourse, not as active subjects. For instance, in the Lapland tourism strategy, it states that in terms of marketing:

> Cooperation brings new and interesting content and interesting meeting opportunities. Versatile content is needed for targeting image marketing locals, potential tourism workers and investors, in addition to the tourists. (RCL, 2015: 23)

In terms of inner and outer possibilities (the possibility to impact one's own future and circumstances), the strategies seem to have an inbuilt idea that the different stakeholders have a similar possibility to take part in the strategy process. For instance, the strategy of Madeira stresses the importance of reflecting the strategy with the 'will and commitment of all stakeholders in sustainable tourism development' (SRETC, 2016: 64). This would mean that people are active and participative in nature and that they want to be part of (sustainable) tourism development. The idea that the possibilities of participation are similar for all is a somewhat Western–liberal notion in the sense that it does not take into consideration the different backgrounds of the expected participants. Having said that, it must be acknowledged that all the strategies recognise the importance of different stakeholders participating in development. For example, the Portuguese strategy emphasises the open and participatory strategy process, and invites readers to join the strategy at the end of the document (TDP, 2017: 26, 32). The policy orientation is noteworthy, however: such a view would suggest that the participants are the guests, not the hosts.

Hospitality in the strategy papers is seen as customer service. In that sense, the worldview is very familiar for hospitality industries, which are proud of their traditions, customs and professionalism. Local communities are seen as providers of hospitable services to tourists and, peculiarly enough, it can also be seen that inclusion is partly realised through hospitable attitude. This kind of approach has connotations of openness and honesty, which are respectable values (RCL, 2015: 26). The strategies wish to honour local environments and culture, while offering hospitable services. From this viewpoint, it could be concluded that hospitality is a central part of the worldview in the strategies, defining attitudes towards tourists as well as the quality of tourism services. It is with the same hospitality that the tourism industry invites locals to be a part of tourism development. Indeed, the role of locals is that of guests, and ownership of the development process has not been defined as being possessed by the locals. Could we even question whether the hospitable worldview hinges on or promotes inclusion within the development processes of tourism and hospitality?

Myths, metaphors and deep stories

According to (Inayatullah, 1998: 820), in the fourth layer, 'analysis is at the deep level of a myth or a metaphor. This layer is formed by the unconscious dimensions of the problem, the collective archetypes. The language in this level is less specific. This level aims in "touching the heart instead of reading the head"'.

In this level, we identified the metaphors and myths that lie behind the strategies. These metaphors help to understand the strategies and the issues that lie beneath the obvious content. They help to translate the spirit of the strategies and make it visible.

A myth that could describe the strategies on a general level would be that of the need for continuous growth. In other words, the commonly accepted principles of the Western market economy are the drivers behind the strategies. This is not surprising because tourism strategies are created to support the destinations in navigating the competitive environment. Consequently, the competitive environment is one of the most important tourism development drivers. It is important to recognise the regions that are competitors in the same market and win customers' attention from these markets. The strategy of Lapland proudly states – following Jon Bon Jovi – 'we weren't born to follow' (RCL, 2015: 2). There is a race, and the various regions each want to finish it first and reach the customers. Relating to this, there is the mentality that 'the customer is always right', which in tourism and other service industries is something that is considered to be an elemental part of the industry. Although this sort of straightforward thinking has also started to gain more critical voices in scholarship and in the regions, customers are not welcomed 'as they are'. Considering the well-being of locals has gained ground, along with environmental viewpoints. In older strategies, the focus has not turned towards sustainability, at least not very strongly.

The host–guest relationship in the strategies reflects the myth of feminine care-taking. The purpose of hosts is to take care of customers and their needs. There is a line between guests and hosts – between the front desk and backstage. However, both locals and tourists are invited to the strategy as guests, making the strategy one of the many stages in which tourism takes place. Indeed, interestingly enough, the staged process of tourism continues from the staged tourism experience (see MacCannell, 1973; Cohen, 1988) to staged policymaking and implementation. Equal conversation between different stakeholders might turn out to be superficial if locals are invited to the collaboration as guests. Hospitality towards guests is one of the valued features of the industry, but the host–guest relationship is changing, and this most likely has implications for strategy development in the future.

In terms of participation, the common idea is that equality in tourism exists – that participation is a matter of choice. The 'yes you can' type of attitude can be noted in the strategies, even if it is not articulated as such. Everyone can travel and everyone is able to host. Again, this is an implication of the liberal thought that replicates the common myth concerning the industry. Seen from outside the industry, it might seem that tourism work is desirable and that anyone can perform this kind of work. Skills and knowhow have important roles in the strategies (although this varies between the different ones), but the underlying thought is that tourism brings employment to locals who are willing and able to work in the industry.

At the myth level, it could be concluded that the regions rely on a special notion of hospitality. Furthermore, ideas of cultural differences and special features, even stereotypes, of people in different regions play a part in building the notion of location-specific hospitality (for instance, Finns are quiet and honest; people in rural regions are more hospitable, and so on). Relying on this special kind of hospitality includes the idea that locals, who are the producers of this special kind of hospitality, are also interested in producing tourism.

Hospitality becomes intertwined with the notion of participation in the context of tourism policymaking, creating a set for staged participation. Behind this lies an idea that is common in the Western world: participation can be organised. It can be concluded that this continues the art of staging known in the industry in the context of customer participation.

Conclusion

This chapter presented an analysis concerning the tourism strategies of regions that can be considered peripheral (Finland, Lapland, Portugal, Madeira, Scotland and the Scottish Highlands and Islands). To do this, CLA was used. The purpose of CLA as a method is to provide deeper insight into tourism strategies and to understand how strategies construct present and future tourism development in peripheral regions. As articulated by Inayatullah (1998: 817), language, in the post-structuralist approach, constitutes reality more so than merely symbolising reality.

In general, it can be concluded that the tourism strategies considered in this study are highly homogenous and do not differ from each other significantly. Analysis on the litany level noted that the strategy papers differ in terms of length and timespan, but their content is mostly similar. Tourism is mainly described and measured in terms of numbers and volumes. None of the strategies has a focus on inclusion, even though it is mentioned briefly in some in terms of accessibility or local participation. Neoliberal discourse about sustainability is visible in the general aims and alignments of the strategies.

When examining the strategies on a systemic level, it becomes clear that tourism is seen as detached from other industries. In addition, local people are not seen as an active part of the tourism system. Rather, they are seen as targets of the system: by developing tourism, the regions and people living there benefit and have better prospects. The strategies do not suggest systemic solutions for changing the system towards a more inclusive industry, nor do they describe tourism as an inclusive system.

The worldviews existing behind the strategies are liberal and growth-based: tourism needs to grow and should grow. The strategies are more or less participatory in nature, as the participation of relevant stakeholders is considered an important aspect in all strategies. However, Western liberal and individualistic worldviews also appear in the assumptions about the ability and willingness of different stakeholders to participate in tourism development. Participation is staged in such a way that assumes that everyone has similar possibilities to take part in the strategy process. Meanwhile, the strategy papers themselves designate which relevant stakeholders are to be invited to tourism development. Furthermore, stakeholders are invited by those considered to be tourism experts, such as tourism developers. Thus, invited stakeholders are guests in tourism development, while ownership stays with those leading the process.

The collective archetypes, myths and metaphors behind the strategies are also related to Western ideas of growth and individualism. A common idea in the strategies is that equality in tourism exists and that participation is a matter of choice. 'Yes you can' is a myth that leads to an understanding of participation throughout the strategies. It embraces tourists, local people, employees, tourism companies, authorities, regions and all who are considered as tourism stakeholders. This leads to assumptions about the possibilities created by tourism, desirable tourism work, the regions and companies competing for the number of tourists and, finally, hospitable local people. Interestingly, the staged shape of tourism seems to continue from staged authenticity and staged tourism experiences (MacCannell, 1973; Cohen, 1988; Wang, 1999) to staged policymaking and implementation.

The myth of the participatory nature of tourism development in the strategies could be described by the term *staged participation*. On the one hand, the strategies are staged and ready for people to join and participate; on the other, tourism in general is staged as something innately participative that different stakeholders should, want and can take part in.

It can be assumed that when phenomena such as over-tourism and antagonism among local people towards tourists receive more attention globally, future strategies will not consider participation as self-evident (as the present ones do). It is desirable that the myth of capability, as well as staged participation, will turn into more inclusive worldviews and systems. It is worth pondering how the discussions and processes would change if the issue of staged participation was up-ended. What about accepting that everyone does not have similar possibilities, abilities, resources or desires to participate in tourism development? What about understanding tourism work as a career and an opportunity for self-development, instead of seeing it as a fun job attracting adventurous people to work for one season? These are examples of questions that would merit further consideration when creating alternative futures based on CLA of tourism strategies.

References

Biddulph, R. and Scheyvens, R. (2018) Introducing inclusive tourism. *Tourism Geographies* 20 (4), 583–588.
Cohen, E. (1988) Authenticity and commoditization in tourism. *Annals of Tourism Research* 15 (3), 371–386, doi: 10.1016/0160-7383(88)90028-X10.1016/0160-7383(88)90028-X.
Darcy, S. and Buhalis, D. (2011a) Introduction: From disabled tourists to accessible tourism. In D. Buhalis and S. Darcy (eds) *Accessible Tourism: Concepts and Issues* (pp. 1–20). Bristol: Channel View Publications.
Darcy, S. and Buhalis, D. (2011b) Conceptualising disability. In D. Buhalis and S. Darcy (eds) *Accessible Tourism: Concepts and Issues* (pp. 21–45). Bristol: Channel View Publications.
De Jouvenel, H. (2019) Futuribles: Origins, philosophy, and practices – Anticipation for action. *World Futures Review* 11 (1), 8–18.
Dredge, D., Jenkins, J. and Whitford, M. (2011) Tourism planning and policy: Historical development and contemporary challenges. In D. Dredge and J. Jenkins (eds) *Stories of Practice: Tourism Policy and Planning* (pp. 13–35). Farnham: Ashgate.
Hakkarainen, M. and Tuulentie, S. (2008) Tourism's role in rural development of Finnish Lapland: Interpreting national and regional strategy documents. *Fennia* 186 (1), 3–13.
Hall, C.M. (1994) *Tourism and Politics: Policy, Power and Place*. Chichester: Wiley.
Harju-Myllyaho, A. (2018) Towards accessible hospitality: Intersectional approach to rainbow tourism. Master's thesis, University of Lapland.
Harju-Myllyaho, A. and Jutila, S. (2018) The meaning of accessible tourism. In M. Gorenak and A. Trdina (eds) *Responsible Hospitality: Inclusive, Active, Green* (pp. 9–20) Slovenia: University of Maribor Press.
Harju-Myllyaho, A. and Kyyrä, S. (2013) Megatrendit ja esteetön vieraanvaraisuus. In S. Jutila and H. Ilola (eds) *Matkailua kaikille? Näkökulmia matkailun ennakointiin, osa II* (pp. 8–18). Rovaniemi: Erweko.
HIE (Highlands and Islands Enterprise) (2019) See https://www.hie.co.uk/our-region/ (accessed January 2020).
Höckert, E. (2015) Ethics of hospitality: Participatory tourism encounters in the northern highlands of Nicaragua. PhD dissertation, University of Lapland.
Inayatullah, S. (1990) Deconstructing and reconstructing the future: Predictive, cultural and critical epistemologies. *Futures* 22 (2), 115–141.
Inayatullah, S. (1998) Causal layered analysis: Poststructuralism as a method. *Futures* 30 (8), 815–829.
Inayatullah, S. (ed.) (2004) *The Causal Layered Analysis (CLA) Reader. Theory and Case Studies of an Integrative and Transformative Methodology*. Taipei: Tamkang University Press.

Isola, A.-M., Kaartinen, H., Leeman, L., Lääperi, R., Schneider, T., Valtari, S. and Keto-Tokoi, A. (2017) *Mitä osallisuus on? Osallisuuden viitekehystä rakentamassa*. Working paper. Helsinki: Finnish Institute for Health and Welfare

Jutila, S. and Harju-Myllyaho, A. (2017) Esteettömyys matkailussa. In J. Edelheim and H. Ilola (eds) *Matkailututkimuksen avainkäsitteet* (pp. 223–228). Rovaniemi: Lapland University Press.

Kaboli, A. and Tapio, P. (2017) How late-modern nomads imagine tomorrow? A causal layered analysis practice to explore the images of the future of young adults. *Futures* 96, 36–43.

Kamppinen, M. and Malaska, P. (2003) Mahdolliset maailmat ja niistä tietäminen. In M. Kamppinen, O. Kuusi and S. Söderlund (eds) *Tulevaisuudentutkimus. Perusteet ja sovellukset* (pp. 55–116). Helsinki: Suomen kirjallisuuden seura.

Ketonen-Oksi, S. (2018) Creating a shared narrative: The use of causal layered analysis to explore value co-creation in a novel service ecosystem. *European Journal of Futures Research* 6 (1), 1–12.

MacCannell, D. (1973) Staged authenticity: Arrangements of social space in tourist settings. *American Journal of Sociology* 79, 589–603.

McCabe, S., Minnaert, L. and Diekmann, A. (2012) Introduction. In S. McCabe, L. Minnaert and A. Diekmann (eds) *Social Tourism in Europe: Theory and Practice* (pp. 1–12). Bristol: Channel View Publications.

RCL (Regional Council of Lapland) (2015) Lapland tourism strategy 2015–2018. See https://issuu.com/lapinliitto/docs/lapin_matkailustrategia_2015-2018

Sedgley, D., Pritchard, A. and Morgan, N. (2012) 'Tourism poverty' in affluent societies: voices from inner-city London. *Tourism Management* 33, 951–960.

Selwyn, T. (2000) An anthropology of hospitality. In C. Lashley and A. Morrison (eds) *In Search of Hospitality. Theoretical Perspectives and Debates*. Oxford: Butterworth-Heineman.

Sitra (2019) Images of futures. See https://www.sitra.fi/en/dictionary/images-of-futures/ (accessed March 2020).

Söderlund, S. and Kuusi, O. (2003) Tulevaisuudentutkimuksen historia, nykytila ja tulevaisuus. In M. Kamppinen, O. Kuusi and S. Söderlund (eds) *Tulevaisuudentutkimus. Perusteet ja sovellukset*. Helsinki: Suomen kirjallisuuden seura.

SRETC (Secretaria Regional da Economia, Turismo e Cultura) (2016) Madeira tourism strategy 2017–2021. See http://www.visitmadeira.pt/Admin/Public/Download.aspx?file=Files%2FFiles%2FVisitMadeira%2FEstudos%2Fj-DOCUMENTO-ESTRATEGICO-2017-21.pdf (accessed March 2020).

STA (Scottish Tourism Alliance) (2012) Tourism Scotland 2020. See https://scottishtourismalliance.co.uk/wp-content/uploads/2019/03/Tourism-Scotland-2020-final.pdf (accessed March 2020).

STA (Scottish Tourism Alliance) (2019) Scotland is Now. See https://www.scotland.org/about-scotland/where-is-scotland/map-of-scotland (accessed March 2019).

Statistics Finland (2019) See https://www.tilastokeskus.fi/index_en.html (accessed March 2020).

TDP (Turismo de Portugal) (2016) Estratégia de Turismo Portugal 2027. International focus group English Market. See https://estrategia.turismodeportugal.pt/sites/default/files/letalgarve_files/ET_27_apresentacao_ru_v_completa_fotos_novas.pdf (accessed November 2019).

TDP (Turismo de Portugal) (2017) Portuguese tourism strategy 2027. See https://travelbi.turismodeportugal.pt/en-us/Documents/Strategy/estrategia-turismo-2027.pdf (accessed January 2021).

United Nations (1948) Universal Declaration of Human Rights. See https://www.ohchr.org/EN/UDHR/Documents/UDHR_Translations/eng.pdf (accessed March 2020).

UNWTO (2016) Messages of the World Committee on Tourism Ethics on Accessible Tourism. See https://webunwto.s3.eu-west-1.amazonaws.com/s3fs-public/2019-10/wctemessagesonaccessibletourism.pdf (accessed March 2020).

Vallis, R. and Inayatullah, S. (2016) Policy metaphors: From the tuberculosis crusade to the obesity apocalypse. *Futures* 84, 133–144.

Visit Finland (2019) Finland in facts. See https://finland.fi/facts-stats-and-info/finland-in-facts-2/ (accessed March 2020).

Visit Finland (2010) Finland tourism strategy 2020. See http://www.visitfinland.fi/wp-content/uploads/2013/04/Matkailustrategia_020610.pdf (accessed March 2020).

Visit Portugal (2019) All about Portugal. See https://www.visitportugal.com/en/sobre-portugal/biportugal (accessed March 2020).

Wang, N. (1999) Rethinking authenticity in tourism experience. *Annals of Tourism Research* 26, 349–370, doi:10.1016/S0160-7383(98)00103-0.

4 Where is 'The Poor' in Pro-poor VCA? – A Review of Applying Pro-poor VCA in a Coastal Tourism Destination in the Northeast of Brazil

Theres Winter, Seonyoung Kim and Nicola Palmer

Introduction

Tourism is often considered to be a favourable tool for poverty alleviation and development due to its low entry barrier, contribution to un-/low-skilled employment and spill-over effect to other economic sectors. Over the years, states and international development agencies have explored different tourism-led poverty-reduction approaches with attempts to integrate 'the poor' into the tourism value chain through employment and entrepreneurial activities to increase pro-poor impacts.

The adoption of the Sustainable Development Goals by all United Nations member states in 2015 brought a shift in the international community's focus from pro-poor towards inclusive growth. This transition acknowledges the importance of equality and social justice in poverty-reduction strategies. In the tourism literature, researchers call for the inclusive development of tourism, where marginalised groups involved in tourism production and consumption share the benefits equally (Scheyvens & Biddulph, 2018). This chapter focuses on one such marginalised group – the poor – and attempts to examine how and to what extent pro-poor tourism (PPT) research drives inclusive growth and promotes social sustainability. For this, it is necessary to review the research methodologies and tools currently used and explore more innovative approaches for future studies.

The aims of this chapter are thus to critically evaluate pro-poor value chain analysis (VCA) – a frequently used PPT research method – and to examine the extent to which it contributes to inclusive and socially sustainable tourism development. By analysing a case study in a coastal tourism destination in the northeast of Brazil, the chapter makes recommendations for scholars and practitioners on how to enhance the value of pro-poor VCA as a more inclusive method.

The following literature review illustrates existing perspectives on tourism and poverty reduction, the complexity of poverty and the pro-poor VCA method and its application in tourism research. This is followed by the research methodology for this work, including the study location and the empirical application of the pro-poor VCA method. Following presentation of the key results of the case study, a critical evaluation of pro-poor VCA and its value as an inclusive method is provided. The chapter concludes by highlighting the key findings and offering recommendations.

Literature Review

Perspectives on tourism as a tool for poverty reduction

Since the 1960s, the potential of tourism to reduce poverty has been acknowledged under the neoliberal paradigm that promoted tourism as a tool for development (Sharpley & Telfer, 2015). Generating foreign exchange and creating employment, income and wider social benefits, tourism was considered as 'a catalyst for modernization, economic development and prosperity in emerging nations in the third world' (Williams, 1998: 1). Consequently, many developing countries have applied tourism-led growth strategies (Mowforth & Munt, 2016). However, from the 1970s onwards, it became apparent that tourism was not a panacea for development in developing countries as the multiplier effects were considerably less than expected (Britton, 1982; Brohman, 1996; Mosedale, 2011; Oppermann & Chon, 1997; Scheyvens, 2007). Being sceptical of tourism as a development strategy and its potential to generate foreign exchange, create employment opportunities and stimulate the wider economy, several studies (e.g. Britton, 1982; Brohman, 1996; Mbaiwa, 2005, 2017) have employed Frank's (1969) dependency theory to highlight the structural inequalities in the international tourism industry (Schilcher, 2007). Dependency theorists have argued that economic development through tourism is limited due to high levels of integration, foreign control and ownership, and associated leakages (Britton, 1982; Brohman, 1996). Tourism development has also been considered to create spatial and social inequalities at the destination level since hotel and resort enclaves are established at a distance from local communities and thus interaction between tourists and hosts is not facilitated (Britton, 1982; Mbaiwa, 2005,

2017). In these enclaves, it has been observed that local people have limited opportunities for unskilled and low-paid jobs and they lack the knowledge and skills required for managerial positions (Daldeniz & Hampton, 2013; Mbaiwa, 2017; Mutimucuio & Meyer, 2011).

To overcome existing inequalities in the international mainstream tourism industry, small-scale tourism development as an alternative to large-scale tourism was advocated (Harrison, 2008; Knight *et al.*, 2017). In particular, community-based tourism, with locals being directly involved in tourism planning and decision-making and having direct control over development and benefits, was promoted from the 1990s (Nair & Hamzah, 2015; Sebele, 2010). However, the success of the concept has been questioned, with a range of community-based tourism initiatives having failed in that they are not economically viable due to lack of knowledge, skills and market access (Goodwin & Santilli, 2009; Mitchell & Muckosy, 2008). In response, at the end of the 1990s, the practitioner movement of PPT emerged, seeking to put 'poverty at the heart of the tourism agenda' (Ashley *et al.*, 2000: 1) and achieve 'tourism that generates net benefits for the poor' (Ashley *et al.*, 2001: 2). The central idea of PPT is to develop strategies that seek to integrate poor people in the mainstream tourism industry by advancing employment and entrepreneurial opportunities (Harrison, 2008). However, PPT research focuses on the economic impacts of tourism at the micro level (Rylance & Spenceley, 2017) and pays less attention to the distribution of tourism impacts (Chock *et al.*, 2007), treating tourism as 'pro-poor' even 'if richer people benefit more than poorer people' (Ashley *et al.*, 2001: 2).

The use of income as the key, often the sole, indicator to measure tourism impacts has been criticised by several authors (Bianchi, 2018; Schilcher, 2007; Spenceley & Meyer, 2012) as the wider social, cultural and environmental impacts of tourism on poverty reduction are neglected. Several scholars have thus called for more holistic and broader approaches to the study of tourism and poverty reduction (Bianchi, 2018; Erskine & Meyer, 2012; Meyer, 2009; Scheyvens, 2011; Schilcher, 2007). More recently, tourism scholars have highlighted the need to link poverty and inequality and to incorporate the distribution of tourism impacts into poverty analyses (Bianchi, 2018; Bwalya-Umar & Mubanga, 2018; Truong *et al.*, 2014). These calls for broader approaches to study tourism and poverty and the focus on distributional justice are closely associated with the concept of inclusive tourism, in which the ethical production of tourism and the sharing of tourism benefits is emphasised, for example through the upskilling of local people for quality jobs, the involvement of marginalised groups in tourism decision-making and the improvement of quality of life for local communities (Scheyvens & Biddulph, 2018). Rather than simply focusing on tourism's contribution to job creation and monetary impact, inclusive tourism promotes social sustainability and equality in tourism development, which has been arguably lacking in PPT.

Understanding and researching poverty

Poverty is a complex phenomenon that means different things to different people (Spicker, 2007). Accordingly, defining poverty has created long-standing debate and discussion (Shildrick & Rucell, 2015), and absolute and relative perspectives of poverty have been central (Lister, 2004) (see Table 4.1).

Absolute poverty – a condition in which people cannot afford a minimum standard of living (United Nations, 1995) – is frequently found in developing countries. Relative poverty also occurs in the developed world (United Nations, 1995). From a relative perspective, poverty is defined by social norms and therefore poverty may actually differ depending on the societal context (Townsend, 1979). Moreover, it may be asserted that relative poverty reflects inequality in a society (Spicker, 2007). People are not just poor, but poverty is shaped by conditions in society in which resources are unequally distributed. Poverty and inequality are strongly interrelated: reducing structural inequalities is crucial for tackling poverty effectively (Atkinson, 1999; Hick, 2012; McKay, 2002; Sen, 1995).

To reconcile both definitions of absolute and relative poverty, the United Nations (1995: 38) adopted the definition of overall poverty (see Table 4.1). The definition of overall poverty is comprehensive in that it addresses material conditions/needs (e.g. food, housing), economic circumstances (e.g. resources to obtain livelihood) and social conditions (e.g. participation in social life). Accordingly, poverty is positioned as a multi-dimensional concept and in order to develop understanding of poverty requires taking all dimensions into account (Spicker, 2007).

Table 4.1 Definitions of poverty

Absolute Poverty United Nations (1998: 38)	"a condition characterized by severe deprivation of basic human needs, including food, safe drinking water, sanitation facilities, health, shelter, education and information"
Relative poverty Townsend (1979: 31)	"Individuals, families and groups in the population can be said to be in poverty when they lack the resources to obtain the types of diet, participate in the activities and have the living conditions and amenities which are customary, or are at least widely encouraged or approved, in the societies to which they belong. Their resources are so seriously below those commanded by the average individual or family that they are, in effect, excluded from ordinary living patterns and activities"
Overall Poverty United Nations (1998: 38)	"lack of income and productive resources sufficient to ensure sustainable livelihoods; hunger and malnutrition; ill health; limited or lack of access to education and other basic services; increased morbidity and mortality from illness; homelessness and inadequate housing; unsafe environments; and social discrimination on and exclusion. It is also characterized by a lack of participation in decision-making and in civil, social and cultural life"

To develop understanding of poverty, both quantitative and qualitative approaches have been applied. To date, applications of quantitative measurements have been dominant in poverty research *per se* and in PPT research in particular. One reason for this is that quantitative measurement enables large-scale poverty analyses and the development of intervention strategies to alleviate poverty (Lister, 2004; Spicker, 2007). Indeed, quantitative measurement is based on indicators (e.g. income, consumption or wealth) and allows for assessment of performance against a defined standard; however, the indicator(s) chosen has implications for a study's conclusions or outcomes (Atkinson, 1983; Gallo, 2002). For example, a commonly used indicator is income – a person is considered 'poor' when their generated income is not sufficient to ensure a minimum standard of living (Fukuda-Parr, 2006; Townsend, 2006). This minimum standard is described/prescribed by the notion of the poverty line. The World Bank (2017a), for example, has defined extreme poverty as living on less than US$1.25 per day (updated from 2015 to living on less than US$1.90 per day).

There has been long-standing debate among academics and practitioners about the application of poverty lines to define 'the poor' and measure poverty; critics argue that focusing on income implies that human well-being solely depends on material needs (Townsend, 2006) and this represents only one dimension of poverty (Edward, 2006). In order to acknowledge and represent a multidimensional view of poverty, composite measures of poverty, such as the Multidimensional Poverty Index, have more recently gained support. However, these composite measures have also not avoided criticism – they have been challenged for simplifying the complexity of the problem in that they only produce a numerical indicator and this is considered to be insufficient for capturing a full picture of poverty (Alkire & Santos, 2009; Ranis *et al.*, 2006; Spicker, 2007). Quantitative measures provide crucial information at aggregate level; however, these are able 'to tell only a partial story' (Narayan, 1999: 15). Thus, reliance solely on these measurements risks ineffective design and implementation of policies to fight poverty.

While quantitative approaches seek to objectify poverty through numeric measurement, qualitative approaches are concerned with explaining and developing in-depth understanding of poverty. Qualitative research often utilises a participatory approach as a means of engaging poor people to explore their views on poverty (Lister, 2004). Thus, in theory, qualitative approaches can provide valuable insights into poverty (McGee & Brook, 2001) that can enhance the development and implementation of policies and strategies that really matter to people (Robb, 2002). However, on the other hand, qualitative approaches have limitations; in particular, they do not allow for large-scale intervention strategies to alleviate poverty.

Pro-poor VCA and its application in tourism research

In general, a value chain describes the 'full range of activities to bring a product or service [...] to final consumers' (Kaplinsky, 2000: 121). Tourism, however, has been noted to be:

> no single product but rather a wide range of products and services that interact to provide an opportunity to fulfil a tourist experience that comprises both tangible parts (e.g., hotel, restaurant, or air carrier) and intangible parts (e.g., sunset, scenery, mood). (Judd, 2006: 325)

The tourism value chain is therefore different from conventional value chains due to the complex and fragmented nature of tourism that incorporates planning, transport, accommodation, food and drinks, and shopping and entertainment/activities provided by direct service providers and also depends on secondary supply such as agriculture or construction (see Figure 4.1) (Ashley & Mitchell, 2008).

Furthermore, a tourism value chain is distinct in that the consumer (the tourist) travels to the product (the destination), with production and consumption taking place simultaneously (Ashley & Mitchell, 2008). Due to its complexity and the simultaneity of production and consumption, it

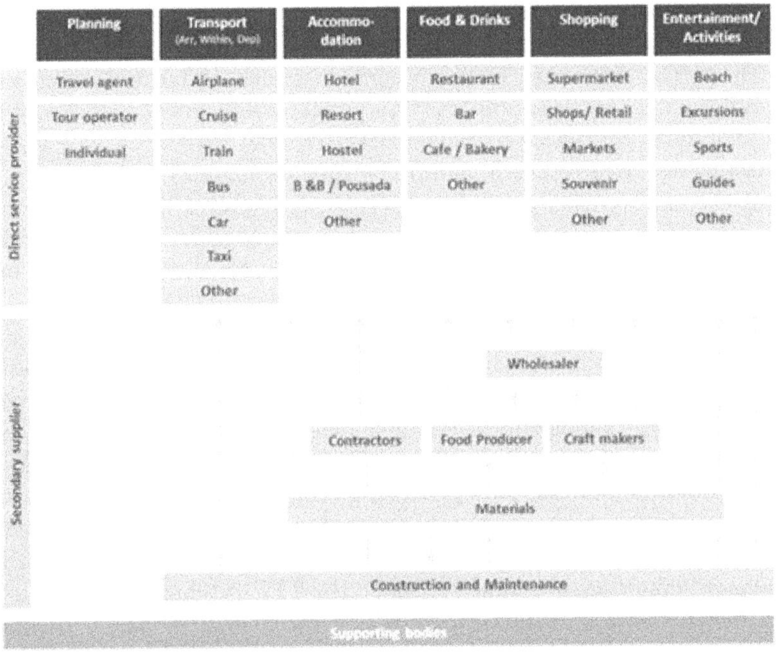

	Planning	Transport (Arr, Within, Dep)	Accommodation	Food & Drinks	Shopping	Entertainment/ Activities
Direct service provider	Travel agent	Airplane	Hotel	Restaurant	Supermarket	Beach
	Tour operator	Cruise	Resort	Bar	Shops/ Retail	Excursions
	Individual	Train	Hostel	Cafe / Bakery	Markets	Sports
		Bus	B &B / Pousada	Other	Souvenir	Guides
		Car	Other		Other	Other
		Taxi				
		Other				
Secondary supplier				Wholesaler		
			Contractors	Food Producer	Craft makers	
				Materials		
			Construction and Maintenance			
			Supporting bodies			

Figure 4.1 Tourism value chain
Source: Authors, adapted from SNV and UNWTO (2010)

may be contended that tourism development provides diverse opportunities for poor people to participate in the value chain (Ashley & Mitchell, 2008).

Mitchell and Ashley (2010) suggest three pathways by which tourism can transfer benefits and costs to local communities: (1) direct impacts, (2) indirect and induced impacts and (3) dynamic effects. The first pathway refers to direct flows from tourism service providers to poor-income groups (e.g. wages from work in the accommodation or food and beverage sectors). The second pathway relates to indirect flows in that tourist activity stimulates secondary sectors, such as the supply of food and beverages or the construction of facilities. It has been argued that the indirect impacts of tourism on poor communities might be significantly higher than direct impacts (Meyer, 2007; Mitchell, 2012). However, secondary products are often not available locally – they need to be sourced from outside the destination, leading to leakages (Mitchell & Ashley, 2010). Induced impacts result from the people employed in tourism spending in the local economy (Mitchell & Ashley, 2010), thus producing a multiplier effect. The third pathway – the dynamic effects of tourist activity – include longer-term non-monetary impacts, for example improved skills due to employment and training in tourism or scarcity of resources (Mitchell & Ashley, 2010).

Given the variety of potential impacts that tourism can have on poverty reduction, it is not surprising that international development agencies such as the Overseas Development Institute and the Netherlands Development Organisation SNV frequently adopt tourism as a tool for development (Spenceley & Meyer, 2012; Truong, 2018). Pro-poor VCA is a commonly used research method by such agencies as it allows them to map the local tourism industry and its actors, to track revenue flows and income, to analyse how much of tourism income reaches 'the poor' and to identify nodes of intervention along the value chain to increase the pro-poor impact (Mitchell & Ashley, 2010; Mitchell *et al.*, 2015).

The availability of a structured framework and practical tools, such as the *Opportunity Study Guidelines* published by the International Trade Centre (ITC), makes the application of pro-poor VCA relatively easy and time- and cost-efficient (Ashley *et al.*, 2009). The ITC, an executing agency of the United Nations Development Programme, seeks to 'link poor communities with promising products and services to markets using technical support, in order to achieve a direct impact on their economic development' (Ashley *et al.*, 2009: 1) through its tourism-led poverty-reduction programme. Since the ITC research tool has been applied, tested and recommended by established practitioners in the field, the pro-poor VCA research design has undergone rigorous scrutiny. Additionally, a standard tool enables greater consistency in conducting assessments, which, in return, produces more reliable and valid results (Ashley *et al.*, 2009).

The ITC *Opportunity Study Guidelines* provide a structured framework based on three phases (see Table 4.2). This chapter focuses on Phase I (Diagnosis of current situation and context), which uses pro-poor VCA to measure tourism-related revenue flows and income.

Phase 1 consists of five steps and several activities for the application of pro-poor VCA, as shown in Table 4.3. Following these steps and activities, the chapter next explains how pro-poor VCA was applied for the case study of a coastal tourism destination in the northeast of Brazil.

Table 4.2 Pro-poor VCA phases

Phase 1	Phase 2	Phase 3
Diagnosis of current situation and context: Map the tourism value chain in order to develop understanding of tourism in the destination and monetary flows	Project opportunities, prioritisation and feasibility: Develop a list of potential interventions in order to increase the pro-poor impact	Project planning: Plan, implement and monitor interventions

Source: Authors, based on Ashley *et al.* (2009)

Table 4.3 Pro-poor VCA Phase 1: Steps and activities

Phase 1			
Step		Activity	
I	Prepare the study	a	Define the scope of the destination
		b	Define the target group(s)
II	Map the big picture i.e. enterprise and other actors in the tourism sector, links between them, demand and supply data, and the pertinent context	c	Get on top of existing information
		d	Identify relevant policies and plans
		e	Map actors along the tourism value chain
III	Map where the target groups do and do not participate	f	Annotate the map to show where the target groups participate
IV	Conduct fieldwork with service providers of each node of the chain and with tourists	g	Tourism value chain survey - gathering information for each node of the value chain
		h	Tourist survey - gathering information from tourists
V	Track revenue flows and income and estimate how expenditure flows through the chain and how much accrues to the target groups	i	Analyse the data and interpret the results
		j	Estimate how much tourism expenditure reaches the target groups via different nodes in the chain

Source: Authors, based on Ashley *et al.* (2009: 17–18)

Methodology

Study location

The case study is a coastal tourist destination in the municipality of Mata de São João in the state of Bahia in the northeast of Brazil. This destination is part of the tourist region Costa dos Coqueiros, which is one of 13 tourist regions in the state of Bahia (SETUR, 2011). The whole region of Costa dos Coqueiros benefitted hugely from a tourism-related investment programme called PRODETUR-NE (Programa de Desenvolvimento do Turismo). This programme was launched in the 1990s by the Brazilian government in cooperation with local governments, international and domestic financial institutions (Inter-American Development Bank and Banco do Nordeste do Brasil) and tourism agencies with the aim of drawing the northern regions out of 'economic backwardness' and promoting social and equitable development (Siegel & Alwang, 2005). The key objective was to reduce poverty through employment in tourism and realise wider economic, social and environmental objectives (e.g. transport, sanitation, education, conservation) by using domestic tourism as a tool for transferring wealth from the richer south to the poorer north (Pegas *et al*., 2015; Siegel & Alwang, 2005). Focusing on mass tourism development along the coastline in the northeast, the first stage of the programme (from 1994–2004) included an investment of approximately US$700 million, which was distributed to economically deprived areas with tourism potential (IADB, 2017; Pegas *et al*., 2015). The second stage, PRODETUR-NE II, included investments of approximately US$400 million and was completed in 2014 (IADB, 2017).

The majority of Brazil's coastal destinations with their sun, sea and sand image are located in the northeast (Pegas *et al*., 2015). The state of Bahia has become particularly popular among travellers. Nine million tourist arrivals were registered in Bahia in 2008, 95% of which were domestic travellers (SETUR, 2011). In 2014, Bahia received approximately 14.5 million tourists (again, 95% being domestic tourists), which shows the extensive growth of tourism in the area (OTB, 2017). The number of domestic travellers in the tourist region Costa dos Coqueiros increased from 450,000 in 2008 (SETUR, 2011) to 1.26 million in 2011 (OTB, 2017).

The latest tourism strategy for Bahia 2007–2016 (SETUR, 2011) expressed the importance of tourism as an economic sector for the state of Bahia. The Bahian state government intended to improve tourist experiences, strengthen the involvement of local communities, protect the natural and cultural heritage and set an example for sustainable development (SETUR, 2011). Tourism has become the most important economic sector in Brazil's northeast and is an important source of income for coastal communities, particularly in Bahia (Duncan, 2013). For example, in the municipality of Mata de São João (in the state of Bahia), more than 50% of the population works in tourism (IBGE, 2011, as cited in Pegas *et al*., 2015).

However, despite the effort to address poverty through tourism development in Brazil's northeast, poverty continues to persist (IPCIG, 2016). Indeed, while most Brazilian municipalities have a poverty rate below 15%, many municipalities in the north and northeast have poverty rates higher than 60%, sometimes up to 90% (IPCIG, 2016). Poverty not only persists in specific geographical regions, but also among racial groups. Notably, poverty rates among Afro-Brazilians are double of those of white Brazilians (Pereira, 2016). Pereira (2016) concluded that, despite reducing overall poverty, the racial division in Brazil between whites and non-whites remains considerable. Similarly, Bucciferro (2017: 191) attests that the situation of non-white people 'is better than it was a few decades ago, and race has become imbued with pride; yet progress has been irregular and opportunity remains far from equal'. Hochstetler noted:

> ... whether the indicator is inequality, income level and depth of poverty, education, life expectancy, or practically any other known measure of development and well-being, the same Brazilians are still clustered in the lowest categories: those of are rural, dark-skinned, and/or in the north and northeastern regions of the country. (Hochstetler, 2010: 5)

The next section outlines how pro-poor VCA was applied in this case study, detailing the study preparation, data collection and data analysis processes. They are explained against the steps and activities identified in Table 4.3.

Application of pro-poor VCA

Study preparation: Activities a–d

In preparation for the study, the geographical scope of the destination was defined by the local borders, using a tourist map for guidance (activity a).

The poverty line of US$5.50 per day per person was applied to define the poor-income group (activity b) according to the World Bank's guidelines for upper–middle income countries such as Brazil (World Bank, 2017a, 2017b, 2018). It needs to be acknowledged that choosing a poverty line is not an easy task since different poverty lines can be applied that will have implications for the whole study process (Ashley et al., 2009). Thus, with effects on the whole study process, a poverty line should be chosen cautiously (Ashley et al., 2009). An analysis of existing information about tourism supply and demand (activity c) and an assessment of available tourism policies and plans (activity d) provided essential contextual knowledge for conducting pro-poor VCA.

Data collection: Activities e–h

Conducting pro-poor VCA was separated by two field-visits – activities e and f (see Table 4.3) were conducted during a pre-visit in September 2016

Table 4.4 Local tourism value chain

	Accommodation	Food & Drinks	Shopping	Entertainment/Activities
Direct service provider	Hotel	Bar [Just drinks]	Craft stall	Beach
	Pousada	Barraca [Beach bar]	Individual seller [Craft]	Sports [Surfing, Riding, etc.]
		Individual seller [Food & drinks]	Shops	Massage
		Restaurant [Lunch & dinner]		Waterfall
		Restaurant [Dinner]		Jangada [Boat trip]

Source: Authors

and activities g and h were carried out in February and March 2017. To map actors along the tourism value chain (activity e) and show where 'the poor' participate (activity f), practitioners often use a participatory approach based on a workshop format at the beginning of conducting VCA (Ashley *et al.*, 2009). While a workshop provides the benefit of getting to know the tourism actors and laying 'the ground for good participation throughout the work', it can be difficult to describe a 'complicated reality' with a large group of participants (Ashley *et al.*, 2009: 23–24). Instead of using a workshop format, in this study the pre-visit was used to identify tourism businesses and actors in the destination and to recognise where poor people participate through the use of available secondary material (e.g. tourism maps) and informal conversations. Although the pre-visit incurred higher costs and prolonged the process, it was considered to be more effective than a workshop in gathering the necessary information. It would have been difficult to manage group dynamics in a workshop format due to the diversity of stakeholders (e.g. in terms of background, nationality, education) and power relations among them. The outcomes of the pre-visit were a high-level stakeholder map and a detailed list of tourism businesses/actors, from which a sample for conducting the tourism value chain survey (activity g) was drawn (Table 4.4).

The tourism value chain survey (activity g) sought to gather information about finance (e.g. rates charged, occupancy, payments to staff and suppliers), tourism supply and the wider sectors (e.g. shopping facilities), tourism demand (e.g. market segments) and impact (e.g. potential linkages) from tourism actors (i.e. business owners and managers) across the value chain. This study followed the associated guidelines for conducting the survey and utilised the available questionnaires for each sector of the tourism value chain, which were contextualised to the case study. Contextualisation appeared to be necessary to reflect the destination's tourism industry and the complex relations of its components and, ultimately, to gather relevant data.

To conduct the tourism value chain survey, a sample was selected from the list of tourism businesses/actors that was a result of the pre-visit to the destination, using stratified random sampling. Dividing the population of tourism actors into relevant strata was crucial for drawing a sample due to the diversity of actors in the tourism value chain. Since the *Opportunity Study Guidelines* (Ashley et al., 2009) do not specifically address sampling, practitioners were contacted and a minimum quota of 50% of each stratum was suggested. Overall, 57 questionnaire surveys were conducted. Applying the ITC's guidelines was beneficial for enabling the involvement of the tourism stakeholders of the destination. By emphasising that the method was established and commonly applied in development studies around the world, participants appeared to be more willing to get involved in the survey. Credibility for the research method was thus gained.

The questionnaire surveys were administered by the first author, which enabled clarification of questions in order to ensure understanding (Gill & Johnson, 2010). King and Horrocks (2011) suggest choosing an appropriate environment, especially when dealing with sensitive information, in order to maintain reliability of data. Since the information to be gathered from the tourism business owners and managers was potentially sensitive (e.g. cost estimates and salaries), the surveys were carried at the participant's work environment where they felt comfortable, while also ensuring the researcher's health and safety.

Due to a lack of tourist statistics for the destination, a tourist survey was also carried out (activity h). The aim of the survey was to gather information about tourists (e.g. nationality, length of stay, purpose of travel) and about their expenditure in the destination (Ashley et al., 2009). To conduct the tourist survey, a self-administered approach was chosen since only descriptive information needed to be gathered in a limited timeframe. Questionnaires were distributed to accommodation providers, who displayed the questionnaires at reception and/or in rooms. Unfortunately, the response rate and quality of the completed questionnaires were relatively low. Therefore, the method was changed to an interviewer-administered, face-to-face approach, conducted using convenience sampling (i.e. potential participants were approached at the beach while they waited for food and/or drinks). Overall, a sample of 75 responses was collected. The data collected was triangulated with data gathered as part of the tourism value chain survey, for example, in terms of market segments.

Data analysis: Activities i and j

The collected data were entered into pre-defined Microsoft Excel based frameworks for analysis (activity i). After entering the data from the tourist survey into the framework, tourist spending for each node (accommodation, food and drinks, shopping, entertainment/activities) was calculated. In terms of the tourism value chain survey, revenue for each node was calculated. Within each node, the average revenue was calculated using the arithmetic mean for different strata. The average revenue per business in each stratum was calculated using the sample data and then extrapolated for the whole population. This process was followed for the different sectors (i.e. shopping, entertainment/activities and other). In terms of the accommodation sector, room prices for each businesses that were not part of the sample were researched (e.g. from websites) and the average occupancy rate (based on the sample) was used to calculate the average revenue for these. Combining all revenues of each node, the total turnover of the local tourism industry was calculated. To estimate the amount reaching the defined target group, income and salary were assessed against the chosen poverty lines within each stratum of a node.

Limitations of applying pro-poor VCA

The limitations of applying pro-poor VCA in this study are associated with the weaknesses of the ITC *Opportunity Study Guidelines* (Ashley et al., 2009).

Firstly, the guidelines do not provide clear instructions on sampling techniques and sizes. In this study, the researcher identified sampling techniques and sizes for the tourism value chain survey after consultation with practitioners. Although the samples were selected according to the standards in practice, the sample sizes still seemed to be rather small to draw valid conclusions.

Secondly, the standard VCA questionnaires do not provide clear definitions of specific terms and concepts (e.g. skilled, semi-skilled and unskilled workforces). In this empirical study, it became apparent that people interpret such terms in different ways. While some managers considered their employees to be 'skilled' because they had been trained on the job since joining the company, others described them as 'unskilled' because they did not have a 'proper' education in school or university. In this study, consistency could be facilitated by the researcher; however, a lack of clarity in the definitions of such key terms and concepts risks inconsistency in the application of pro-poor VCA.

Thirdly, there is only limited guidance provided by the ITC guidelines on how to calculate the economic impacts, with the consequence that different approaches of calculation and different levels of rigour might be applied. These weaknesses influence the validity and reliability of results and, ultimately, compromise the comparability of PPT studies. Thus, the

ITC's promise of providing 'consistency to the way assessments are conducted which will enhance their reliability and validity' (Ashley *et al.*, 2009: 6) should be treated with caution.

Fourthly, it can be difficult to collect the necessary financial information required for a pro-poor VCA study (Rylance & Spenceley, 2017). There is the possibility that research participants might not report accurate financial information. For example, participants might communicate smaller revenue figures to avoid tax increases or report a higher wage level when employees actually receive less than the minimum salary to avoid a penalty for violating labour laws and regulations. This would further weaken the reliability of pro-poor VCA data. In this empirical study, however, being an outsider to the local tourism industry supported the researcher's integrity in keeping the information confidential. This experience contrasts with Bulmer's (1993) argument that, particularly in developing countries, issues of trust are frequently connected with suspicion towards researchers 'from outside' and a potential misuse of information. Since the participants frequently expressed a high level of mistrust towards public institutions, it seemed that being an outsider and having no affiliation within the destination area was advantageous, as also acknowledged in the ITC guidelines (Ashley *et al.*, 2009). In fact, many participants asked directly who and which institutions had access to the data and would only release sensitive data information (e.g. financial overviews) after reassurance that the data was treated confidentially in line with the university's ethics policy.

It also became evident that record-keeping proved a challenge for small-sized enterprises and micro-entrepreneurs. Many participants who were self-employed simply did not know the occupancy rate of their pousada (Brazilian type of accommodation) or the average number of dishes they served per day in their restaurants or beach bars; they reportedly did not care about these details as long as they had some money left at the end of the day. Since pro-poor VCA is reliant on systematic and accurate records and data, a lack thereof has implications for the reliability of results.

The next section presents the key results of the pro-poor VCA analysis of the coastal tourist destination in the northeast of Brazil. It draws on the empirical work and focuses on three aspects of the pro-poor VCA results: (1) local tourism value chain, (2) tourism demand and (3) local tourism-related revenue flows and income.

Pro-poor VCA Results

Local tourism value chain

The local tourism value chain was found to encompass four nodes: accommodation; food and drinks; shopping; entertainment/activities. In each sector, different direct service providers were mapped (see Table 4.4).

Local transport (i.e. motorcycle-taxi) is available in the area, but was not reported to be used by tourists due to the small size of the destination. In addition, local secondary suppliers (e.g. agriculture) in the destination were not identified because products and materials were reportedly sourced externally to the local area, mainly from Bahia's capital city Salvador. This was often due to the low quality of products combined with a poor and unreliable service of local suppliers.

In terms of accommodation, 27 pousadas and 2 hotels offering a wide range of quality (from basic to luxurious) were mapped. On average, the pousadas had 36 beds and were thus much smaller than both hotels. The average price for a pousada was BR$273 (= GBP66) for a double bedroom per night. The prices varied greatly depending on low season or high season (Christmas and New Year up to carnival in February). For example, during Christmas and New Year, prices can be three times higher than in July/August. As a result of the very short season, the yearly average occupancy rate for pousadas was only 40%. In contrast to pousadas, both hotels reported an occupancy rate of 80% throughout the year, and at much higher daily rates. Pousadas offered bed and breakfast; further services such as a bar, tour booking and airport transfer were only available at extra cost. Accordingly, the revenue of the pousadas was mainly generated from renting rooms – on average, 90% of revenue was generated from rooms and 10% from food and drinks. One of the hotels in the study was a luxurious all-inclusive hotel and resort offering all the extra services a guest may wish for (e.g. candlelit dinner at the beach), while the second offered mainly bed and breakfast and a variety of services and activities for extra cost.

In terms of the food and drinks node, five main types were identified (see Table 4.4). Overall, there were 60 providers of food and drinks. The bars were medium-scale (approximately 30 chairs) and, located in the centre of the village, visited by locals and tourists alike. These businesses were family-run, meaning they did not have an extra impact on local employment. The barracas (beach bars) were found to have become largely standardised in terms of size, food and drink offerings and prices, and service. Furthermore, tourists were able to buy products from individual sellers at the beach (beach hawkers) selling ice cream, home-made sweets and nuts, cheese and *acarajé* (a traditional dish in the state of Bahia). The municipality intended to limit the growth of this informal market by requiring a registration from all hawkers, which involved a small annual fee. The Brazilian cuisine was found to be for lunch rather than dinner, and thus typical Brazilian/Bahian restaurants were open for lunch but not all establishments were open for dinner, depending on the season. The Brazilian/Bahian restaurants varied in size (from 25 to 100 chairs) and sales to tourists rather than locals (50–80% to tourists). Notably, the restaurants that solely offered dinner had 30–80 chairs and 70–80% of sales to tourists.

With regard to shopping, about six shops sold clothes (mainly beachwear) and souvenirs, and three stalls sold crafts, with each selling different products (wooden, bast fibre and metallic products). Tourists could also buy souvenirs (e.g. bracelets, necklaces, earrings) from individual sellers (beach hawkers). To sell such products (mainly at the beach), the hawkers needed to be registered with the municipality. All their revenue was generated from tourism – thus, without tourism, the individual sellers would not have any income.

Offers of activities for tourists were very limited and tourists often felt that they need to take the initiative to learn about what is offered rather than being getting actively directed towards activities.

Tourism demand

The tourist survey identified that tourism demand largely builds on the leisure market, with 73% of participants indicating that they travelled to the destination for holidays, 16% to visit friends and relatives and only 11% were on business or travelling for other purposes. This mirrored the image of Brazil's northeast as a typical holiday destination (OTB, 2017). In terms of nationality, 60% of participants were from Brazil and 40% were international travellers (from Germany, Switzerland and Argentina); however, the proportion of Brazilian travellers in terms of tourism demand by nationality was, on average, likely to be higher than at the point of survey, which fell towards the end of the tourist season when the number of domestic tourists decreased. In terms of length of stay, it was found that international travellers stayed for an average of 10 days; this was notably longer than Brazilian tourists who stayed for five days on average.

Tourists staying in pousadas were largely independent travellers, while tourists staying in the large hotel and resort had booked an all-inclusive package. An overview of tourist spending along the tourism value chain (accommodation, food and drinks, shopping and entertainment/activities) is shown in Figure 4.2, presented separately for independent and package tourists. On average, independent tourists paid GBP33 per night for accommodation. Alongside spending on accommodation, independent tourists spent approximately GBP11 per day for food and drinks in the destination – in restaurants or at the beach. Relatively little spending took place for shopping and entertainment/activities – potentially due to the limited offers.

Similarly, package tourists spent very little money on shopping and entertainment/activities outside of their accommodation since they reported that they rarely left the hotel complex to visit the destination centre. This was potentially due to the hotel and resort being designed to function as a village on its own or, in order words, as an enclave (Britton, 1982; Mbaiwa, 2005, 2017). Concerns about safety and security outside

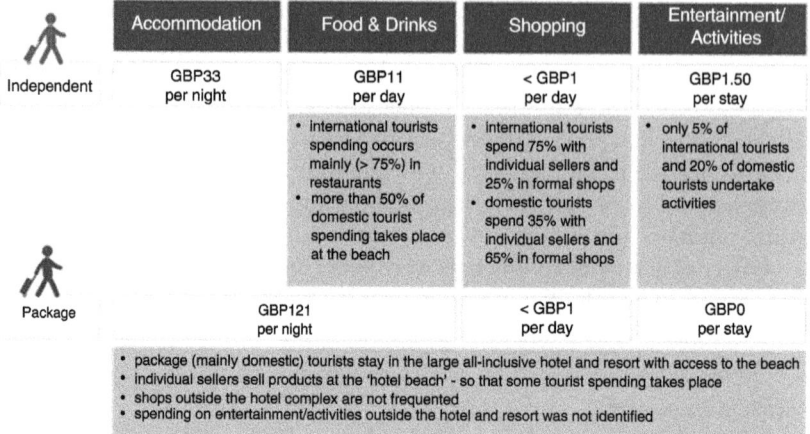

Figure 4.2 Tourist spending
Source: Authors

of the hotel complex and weak transport links between the complex and the village also contributed to tourists not leaving 'the enclave'.

Local tourism-related revenue flows and income

It was estimated that approximately GBP54 million enters the tourism value chain annually, from which GBP52 million (96%) is generated in the accommodation sector (see Figure 4.3). As the hotels were large in terms of the number of rooms and beds *vis-à-vis* the total destination bedspace, had a significantly higher occupancy rate and a significantly higher price (compared with other local accommodation), both hotels together generated 92% of the accommodation revenue. A revenue of GBP2 million (4%) was generated by the three remaining sectors combined (food and drinks, shopping, entertainment/activities). The revenue of GBP1.7 million from food and drinks was the result of spending by independent travellers. As secondary supplies were imported from outside the destination, tourist spending did not remain in the destination.

To examine how much money reached 'the poor', income and salaries were assessed against the poverty line for upper–middle income countries of US$5.50 per day per person (World Bank, 2017a, 2018). The three nodes of food and drinks, shopping and entertainment/activities were found to provide direct, but limited, economic benefits to poor people through micro-entrepreneurship (i.e. individual sellers). Secondary products were mainly sourced from outside the destination and therefore the indirect impacts of tourism did not remain locally.

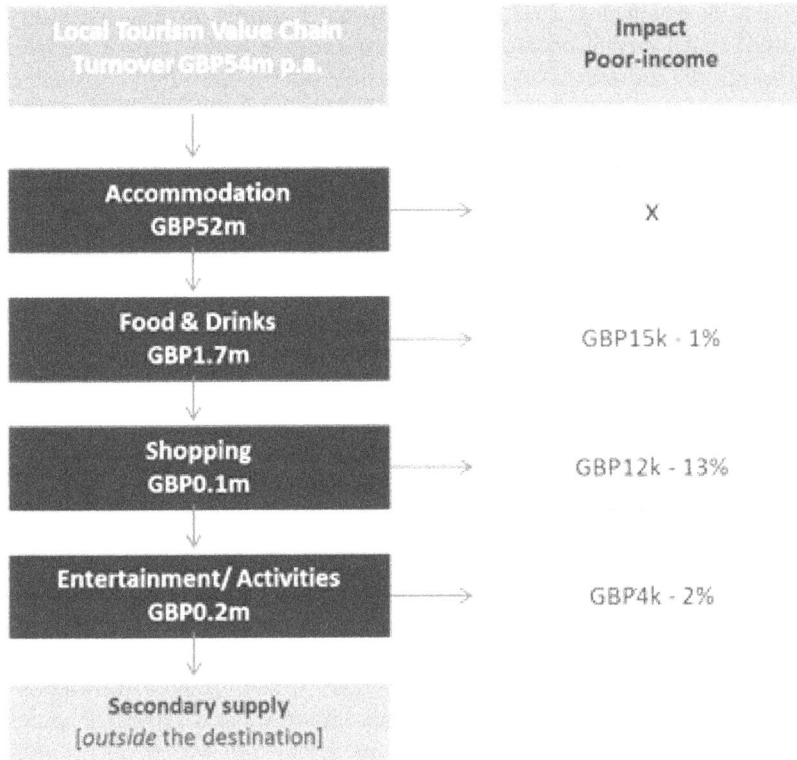

Figure 4.3 Local tourism-related revenue flows and pro-poor income
Source: Authors

Discussion

Pro-poor VCA is a commonly used method in PPT research to map local tourism industries, identify flows of revenue and income, analyse tourism income for poor people and define intervention strategies to increase the pro-poor impact of tourism (Mitchell & Ashley, 2010; Mitchell et al., 2015). Although the validity and reliability of the VCA method is debatable to an extent, the application of pro-poor VCA in this study provided detailed insight into the components of a local tourism industry and identified the involvement of poor people in the value chain. It also generated detailed data on tourism demand (e.g. tourist spending along the value chain), which was not available prior to this research. The improved insight into the local tourism value chain and tourism demand can thus be used as a basis for future intervention strategies to enhance both the tourist experience and the economic impact of tourism on the destination.

Pro-poor VCA focuses on analysing the direct and indirect monetary impacts of tourism on local communities (Mitchell & Ashley, 2010;

Rylance & Spenceley, 2017). The empirical findings of this case study showed that tourism provides direct economic impacts to poor people. Although the indirect economic impacts of tourism are often considered to be more significant than direct impacts (Meyer, 2007; Mitchell, 2012), in this study, the indirect impacts did not occur within the destination since products were mainly sourced from outside the destination, confirming the observation of Mitchell & Ashley (2010). Pro-poor VCA fails to incorporate the induced and dynamic impacts sufficiently, which can both significantly affect the livelihoods of people (Mitchell & Ashley, 2010).

Utilising income as a key indicator for poverty, pro-poor VCA considers a person as poor when the income generated from tourism does not enable a minimum standard of living, which is assessed through the application of poverty lines. This approach, however, does not consider other potential sources of income that people may have or the number of people who may be dependent on the income generated by one person. This has further implications since it is assumed that people's well-being depends only on achieving material needs (Townsend, 2006). Thus, pro-poor VCA represents only one dimension of a multidimensional phenomenon, which simplifies the complexity of poverty. Spicker (2007) argues that, to develop understanding of poverty, all dimensions (e.g. material conditions, economic circumstances and social conditions) need to be taken into account, which is mirrored in the United Nations' (1995) definition of overall poverty. However, pro-poor VCA does not facilitate an analysis of wider dimensions of poverty. In that regard, Spenceley and Meyer (2012: 308) argue that 'perhaps one of the main weaknesses of the VCA approach is its concentration on flows of money, and its lack of attention to environmental, socio-cultural, and political and governance aspects of sustainable development'. Therefore, it can be argued that the narrow focus of pro-poor VCA on analysing income–poverty limits the method's value from becoming a tool for more inclusive development of tourism and promoting social justice.

Since solely relying on income as a key indicator of poverty does not provide a full picture (Alkire & Santos, 2009; Narayan, 1999; Ranis *et al.*, 2006; Spicker, 2007), pro-poor VCA runs the risk of defining ineffective intervention strategies. The fact that pro-poor VCA is mainly conducted with the owners and managers of tourism businesses and excludes the views of people living below the poverty line contributes to the issue of designing and implementing narrow, ineffective strategies. Without including poor people in pro-poor VCA research, the extent to which the method actually contributes to inclusive and socially sustainable tourism development can be questioned.

Through the development of strategies, PPT focuses on increasing net benefits for poor people and, as such, PPT is not concerned with an equal distribution of tourism benefits (Ashley *et al.*, 2001). This is reflected in the method of pro-poor VCA since distributional issues are not considered. However, it has been argued that it is crucial to integrate distributional aspects into poverty analysis since poverty and inequality are strongly interrelated (Atkinson, 1999; Hick, 2012; McKay, 2002; Sen, 1995). The importance of combining both inequality and poverty analysis is particularly the case in the Brazilian context where poverty seems to be strongly linked to racial groups, with poverty rates significantly higher among Afro-Brazilians (Pereira, 2016). This view has become prominent in the tourism literature more recently (Bianchi, 2018; Bwalya-Umar & Mubanga, 2018; Truong et al., 2014), which is reflected in the concept of inclusive tourism. Accordingly, for tourism development to be inclusive, the benefits need to be shared equally among involved groups (Scheyvens & Biddulph, 2018). Therefore, for pro-poor VCA to enable more inclusive and socially sustainable development of tourism, an analysis of the distribution of tourism impacts needs to be integrated.

Conclusion

This chapter examined the extent to which current PPT research drives inclusive growth and promotes social sustainability through a critical evaluation of the pro-poor VCA research method based on a case study of a coastal destination in the northeast of Brazil.

A great strength of pro-poor VCA is that it allows scholars and practitioners to map a particular tourism industry, to examine tourism revenue and income flows and to determine how much tourism income reaches poor people. Thus, it generates data on tourism supply and tourism demand that is often not available, especially in the context of developing countries. Although its application requires more attention to validity and reliability, pro-poor VCA is good starting point for poverty analysis by undertaking an analysis of tourism-related income. It can support the development of strategies that seek to enhance both the tourist experience and the economic impact of tourism.

The current application of pro-poor VCA has three fundamental limitations in promoting more inclusive and socially sustainable development of tourism. Firstly, by using income as a single indicator to analyse poverty, pro-poor VCA ignores the complexity of poverty since it reflects only one dimension of a multidimensional phenomenon. Secondly, pro-poor VCA does not include poor people and their views on poverty. As a consequence of both, pro-poor VCA might actually not lead to the development and implementation of effective pro-poor strategies – which is 'supposedly' the ultimate aim of pro-poor VCA. Thirdly, pro-poor VCA does not integrate an analysis of the distribution of tourism impacts and therefore fails to respond to issues of distributional justice – arguably a key issue in the pursuit of socially sustainable futures.

To overcome the shortcomings of pro-poor VCA and to strengthen its value as a more inclusive method requires not only 'putting poverty at the heart of the tourism agenda' (Ashley *et al.*, 2000: 1), but also putting poor people at the heart of tourism research. By engaging poor people through a participatory approach in the research process, their qualitative perspectives on tourism, poverty and inequality can be gathered, which would complement quantitative data and provide a wider, more inclusive social perspective. In that regard, qualitative research to complement pro-poor VCA should seek to explore the following questions.

- How does income from tourism support people's livelihoods?
- How do the broader impacts (political, social, cultural, environmental) of tourism development influence poverty?
- How are tourism impacts distributed? How does the distribution influence poverty?
- What are potential strategies to fight poverty and inequality?

Using a mixed-methods approach will provide enhanced knowledge and understanding of the complex and multidimensional characteristics of poverty and inequality from the perspectives of people who experience poverty and will, ultimately, support the development of strategies that actually matter to poor people. In this way, research will enable a more inclusive and socially sustainable development of tourism – one where poor people are heard, involved in and equally share the benefits of tourism.

References

Alkire, S. and Santos, J.M. (2009) Poverty and inequality measurement. In S. Deneulin and L. Shahani (eds) *An Introduction to the Human Development and Capability Approach* (pp. 121–143). London: Earthscan.

Ashley, C. and Mitchell, J. (2008) *Doing the Right Thing Approximately not the Wrong Thing Precisely: Challenges of Monitoring Impacts of Pro-poor Interventions in Tourism Value Chains*. ODI Working Paper 291. London: Overseas Development Institute.

Ashley, C., Boyd, C. and Goodwin, H. (2000) Pro-poor tourism: Putting poverty at the heart of the tourism agenda. See https://www.odi.org/sites/odi.org.uk/files/odi-assets/publications-opinion-files/2861.pdf (accessed January 2021).

Ashley, C., Roe, D. and Goodwin, H. (2001) *Pro-poor Tourism Strategies: Making Tourism Work for the Poor. A Review of Experience*. Nottingham: Russell Press.

Ashley, C., Mitchell, J. and Spenceley, A. (2009) *Opportunity Study Guidelines – How to Assess the Potential of a Viable Inclusive Tourism Intervention*. Geneva: International Trade Centre.

Atkinson, A.B. (1983) *The Economies of Inequality* (2nd edn). Oxford: Clarendon Press.

Atkinson, A.B. (1999) The contributions of Amartya Sen to welfare economics. *Scandinavian Journal of Economics* 101 (2), 173–190.

Bianchi, R. (2018) The political economy of tourism development: A critical review. *Annals of Tourism Research* 70, 88–102.

Britton, S. (1982) The political economy of tourism in the third world. *Annals of Tourism Research* 9 (3), 331–358.

Brohman, J. (1996) New directions for tourism in third world development. *Annals of Tourism Research* 23, 48–70.

Bucciferro, J. (2017) Racial inequality in Brazil from independence to the present. In L. Bértola and J. Williamson (eds) *Has Latin American Inequality Changed Direction? Looking over the Long Run* (pp. 171–194). Cham: Springer.

Bulmer, M. (1993) Interviewing and field organisation. In M. Bulmer and D.P. Warwick (eds) *Social Research in Developing Countries: Surveys and Censuses in the Third World* (pp. 212–217). London: Wiley.

Bwalya-Umar, B.H. and Mubanga, K. (2018) Do locals benefit from being in the 'tourist capital'? Views from Livingstone, Zambia. *Tourism and Hospitality Research* 18 (3), 333–345.

Chock, S., Macbeth, J. and Warren, C. (2007) Tourism as a tool for poverty alleviation: A critical analysis of 'pro-poor tourism' and implications for sustainability. In C.M. Hall (ed.) *Pro-poor Tourism: Who Benefits? Perspectives of Tourism and Poverty Reduction* (pp. 34–55). Clevedon: Channel View Publications.

Daldeniz, B. and Hampton, M. (2013) Dive tourism and local communities: Active participation or subject to impacts? Case studies from Malaysia. *International Journal of Tourism Research* 15 (5), 507–520.

Duncan, R. (2013) Tourism market in Brazil. See http://thebrazilbusiness.com/article/tourism-market-in-brazil (accessed April 2016).

Edward, P. (2006) The ethical poverty line: A moral definition of absolute poverty. In D. Ehrenpreis (ed.) *What is Poverty? Concepts and Measures* (pp. 14–16). Brasilia: International Poverty Centre.

Erskine, L.M. and Meyer, D. (2012) Influenced and influential: The role of tour operators and development organisations in tourism and poverty reduction in Ecuador. *Journal of Sustainable Tourism* 20 (3), 339–357.

Frank, A.G. (1969) *Latin America: Underdevelopment or Revolution? Essays on the Development of Underdevelopment and the Immediate Enemy*. New York: Monthly Review Press.

Fukuda-Parr, S. (2006) The human poverty index: A multidimensional measure. In D. Ehrenpreis (ed.) *What is Poverty? Concepts and Measures* (pp. 7–9). Brasilia: International Poverty Centre.

Gallo, C. (2002) *Economic Growth and Income Inequality: Theoretical Background and Empirical Evidence*. Working Paper No. 119. London: Development Planning Unit, University College London.

Gill, J. and Johnson, P. (2010) *Research Methods for Managers* (4th edn). London: Sage.

Goodwin, H. and Santilli, R. (2009) *Community-based Tourism: A Success?* Occasional Paper 11. Leeds: ICRT.

Harrison, D. (2008) Pro-poor tourism: A critique. *Third World Quarterly* 29 (5), 851–868.

Hick, R. (2012) The capability approach: Insights for a new poverty focus. *Journal of Social Policy* 41 (2), 291–308.

Hochstetler, K. (2010) Brazil's GINI coefficient: Can it be beaten? Canada Watch. See Canada_Watch_Fall_2010.pdf (flacso.org.ar)

IADB (Inter-American Development Bank) (2017) Tourism on the rise. See https://www.iadb.org/en/news/webstories/2011-11-01/brazil-prodetur-national-tourism-program%2C9505.html (accessed September 2018).

IPCIG (International Policy Centre for Inclusive Growth) (2016) Poverty profile: The rural north and northeast regions of Brazil. See https://ipcig.org/pub/eng/PRB50_Poverty_profile_the_rural_North_Northeast_regions_of_Brazil.pdf (accessed January 2021).

Judd, D. (2006) Commentary: Tracing the commodity chain of global tourism. *Tourism Geographies* 8 (4), 323–336.

Kaplinsky R. (2000) Spreading the gains from globalisation: What can be learned from value chain analysis? *Journal of Development Studies* 37 (2), 117–146.

King, N. and Horrocks, C. (2011) *Interviews in Qualitative Research*. London: Sage.

Knight, D., Cottrell, S., Pickering, K., Bohren, L. and Bright, A. (2017) Tourism-based development in Cusco, Peru: Comparing national discourses with local realities. *Journal of Sustainable Tourism* 25 (3), 344–361.

Lister, R. (2004) *Poverty*. Oxford: Polity Press.

Mbaiwa, J. (2005) Enclave tourism and its socio-economic impacts in the Okavango Delta, Botswana. *Tourism Management* 26, 157–172.

Mbaiwa, J. (2017) Poverty or riches: Who benefits from the booming tourism industry in Botswana? *Journal of Contemporary African Studies* 35 (1), 93–112.

McGee, R. and Brook, K. (2001) *From Poverty Assessment to Policy Change*. IDS Working Paper No. 113. Brighton: Institute of Development Studies.

McKay, A. (2002) *Defining and Measuring Inequality*. ODI Inequality Briefing Paper No. 1. London: Overseas Development Institute.

Meyer, D. (2007) Pro-poor tourism: From leakages to linkages. A conceptual framework for creating linkages between the accommodation sector and 'poor' neighbouring communities. *Current Issues in Tourism* 10 (6), 558–583.

Meyer, D. (2009) Pro-poor tourism: Is there actually much rhetoric? And, if so, whose? *Tourism Recreation Research* 34 (2), 197–199.

Mitchell, J. (2012) Value chain approaches to assessing the impact of tourism on low-income households in developing countries. *Journal of Sustainable Tourism* 20 (3), 457–475.

Mitchell, J. and Ashley, C. (2010) *Tourism and Poverty Reduction: Pathways to Prosperity*. London: Earthscan.

Mitchell, J. and Muckosy, P. (2008) A misguided quest: Community-based tourism in Latin America. ODI Working Paper No. 102. London: Overseas Development Institute.

Mitchell, J., Font, X. and Li, S. (2015) What is the impact of hotels on local economic development? Applying value chain analysis to individual businesses. *Anatolia* 26 (3), 347–358.

Mosedale, J. (2011) Re-introducing tourism to political economy. In J. Mosedale (ed.) *Political Economy of Tourism: A Critical Perspective* (pp. 1–13). Abingdon: Routledge.

Mowforth, M. and Munt, I. (2016) *Tourism and Sustainability: Development, Globalization and New Tourism in the Third World* (4th edn). Abingdon: Routledge.

Mutimucuio, M. and Meyer, D. (2011) Pro-poor employment and procurement: A tourism value chain analysis of Inhambane peninsula, Mozambique. In R. Dium, D. Meyer, J. Saarinen and K. Zellmer (eds) *New Alliances for Tourism, Conservation and Development in Eastern and Southern Africa* (pp. 27–47). Delft: Eburon.

Nair, V. and Hamzah, A. (2015) Successful community-based tourism approaches for rural destinations: The Asia pacific experience. *Worldwide Hospitality and Tourism Themes* 7 (5), 429–439.

Narayan, D. (1999) *Voices of the Poor – Can Anyone Hear Us?* Washington, DC: World Bank.

Oppermann, M. and Chon, K. (1997) *Tourism in Developing Countries*. London: International Thomson Business Press.

OTB (Observatório do Turismo da Bahia) (2017) Indicatores. See http://observatorio.turismo.ba.gov.br/ (accessed January 2017).

Pegas, F.D.V., Weaver, D. and Castley, G. (2015) Domestic tourism and sustainability in an emerging economy: Brazil's littoral pleasure periphery. *Journal of Sustainable Tourism* 23 (5), 1–22.

Pereira, C. (2016) *Ethno-racial Poverty and Income Inequality in Brazil*. Working Paper No. 60. New Orleans, LA: CEQ Institute.

Ranis, G., Stewart, F. and Samman, E. (2006) Human development: Beyond the HDI. In D. Ehrenpreis (ed.) *What is Poverty? Concepts and Measures* (pp. 12–13). Brasilia: International Poverty Centre.

Robb, C. (2002) *Can the Poor Influence Policy? Participatory Poverty Assessments in the Developing World*. Washington, DC: World Bank.

Rylance, A. and Spenceley, A. (2017) Reducing economic leakages from tourism: A value chain assessment of the tourism industry in Kasane, Botswana. *Development Southern Africa* 34 (3), 295–313.

Scheyvens, R. (2007) Exploring the tourism-poverty nexus. *Current Issues in Tourism* 10 (2–3), 231–254.

Scheyvens, R. (2011) *Tourism and Poverty*. New York: Routledge.

Scheyvens, R. and Biddulph, R. (2018) Inclusive tourism development. *Tourism Geographies* 20 (4), 589–609.

Schilcher, D. (2007) Growth versus equity: The continuum of pro-poor tourism and neo-liberal governance. *Current Issues in Tourism* 10 (2–3), 166–193.

Sebele, L. (2010) Community-based tourism ventures, benefits and challenges: Khama Rhino Sanctuary Trust, Central District, Botswana. *Tourism Management* 31, 136–146.

Sen, A. (1995) *Inequality Reexamined*. Cambridge, MA: Harvard University Press.

SETUR (Secretaria de Turismo do Governo do Estado da Bahia) (2011) Estratégia turística da Bahia 2007|2016. See https://issuu.com/turismobahia/docs/estrat_gia_tur_stica_da_bahia_setur (accessed on January 2017).

Sharpley, R. and Telfer, D.J. (2015) *Tourism and Development: Concepts and Issues* (2nd edn). Bristol: Channel View Publications.

Shildrick, T. and Rucell, J. (2015) *Sociological Perspectives on Poverty*. York: Joseph Rowntree Foundation.

Siegel, P.B. and Alwang, J.R. (2005) *Public Investments in Tourism in the Northeast of Brazil: Does a Poor-area Strategy Benefit the Poor?* Latin America and Caribbean Region Sustainable Development Working Paper 22. Washington, DC: World Bank.

SNV and UNTWO (2010) *Manual on Tourism and Poverty Alleviation: Practical Steps for Destinations*. Madrid: World Tourism Organization.

Spenceley, A. and Meyer, D. (2012) Tourism and poverty reduction: Theory and practice in less economically developed countries. *Journal of Sustainable Tourism* 20 (3), 297–317.

Spicker, P. (2007) *The Idea of Poverty*. Bristol: Policy Press.

Townsend, P. (1979) *Poverty in the United Kingdom: A Survey of Household Resources and Standards of Living*. Harmondsworth: Penguin.

Townsend, P. (2006) What is poverty? An historical perspective. In D. Ehrenpreis (ed.) *What is Poverty? Concepts and Measures* (pp. 5–6). Brasilia: International Poverty Centre.

Truong, V.D. (2018) Tourism, poverty alleviation, and the informal economy: The street vendors of Hanoi, Vietnam. *Tourism Recreation Research* 43 (1), 52–67.

Truong, V.D., Hall, C.M. and Garry, T. (2014) Tourism and poverty alleviation: Perceptions and experiences of poor people in Sapa, Vietnam. *Journal of Sustainable Tourism* 22 (7), 1–19.

United Nations (1995) *The Copenhagen Declaration and Programme of Action, World Summit for Social Development, 6–12 March 1995*. New York: United Nations.

Williams, S. (1998) *Tourism Geography*. London: Routledge.

World Bank (2017a) A richer array of international poverty lines. See http://blogs.worldbank.org/developmenttalk/richer-array-international-poverty-lines (accessed February 2018).

World Bank (2017b) Safeguarding against a reversal in social gains during the economic crisis in Brazil. See http://documents.worldbank.org/curated/en/567101487328295113/pdf/112896-WP-P157875-PUBLIC-ABSTRACT-SENT-SafeguardingBrazilEnglish.pdf/ (accessed September 2018).

World Bank (2018) World Bank economy classification by country. See http://gi.org/wp-content/uploads/2018/01/World-Bank-Country-List-Upper-Middle-and-High.pdf (accessed February 2018).

Part 3
Practices

5 Information about Tourism Destinations' Accessibility in Tourism Online Platforms: Is it Useful for People with Diverse Abilities?

Asunción Fernández-Villarán, Mónica Erice, Nagore Espinosa, Ana Goytia and Aurora Madariaga and Ainara Rodríguez

> To Ainara Rodríguez, our dear friend and colleague, a bright and warm spirit we had the pleasure to encounter in our personal and professional lives. This is one of many pieces of work and contributions you have made for the world to be a better place. We miss you very much.

Introduction

It is estimated that Europe will be home to more than 160 million people with diverse capabilities by 2025, of which 70% will have the economic and physical means to travel (Soriano, 2017). This rapidly growing segment of tourists includes all age groups, persons with temporary and permanent disability and their companions, as well as senior tourists.

Although people with disabilities still travel far less often than people without any disability (Cole *et al.*, 2019), a report from the Spanish Accessible Tourism Observatory (Bowtell, 2015) shows that 50% of these travellers take between two and three trips every year, mainly during the summer months, and usually to sun and beach destinations. The first criteria when choosing a destination is based on its tourism appeal, followed by economic factors (Bowtell, 2015). However, these two criteria are not the only ones that affect functionally diverse visitors when making the final decision of booking the trip; the destination's universal accessibility and correct and reliable access to related information are also understood to be key criteria (Hernández-Galán, 2017; Kołodziejczak, 2019). Specifically, on a scale from 0 to 10, visitors with special needs gave a score of 7.3 to the importance of having information about a destination's accessibility and resources (Bowtell, 2015).

The Spanish Accessible Tourism Observatory report also shows that nearly 7 out of every 10 travellers book direct services online, followed by online booking (11%) and offline (8%) travel agencies, and then through associations associated with their disability (11%) (Bowtell, 2015). It should also be noted that 10% of those surveyed relied on friends or family members to organise and book their travel. These data thus reflect the relevance of the information and opinions shared on forums, blogs, websites and search engines, and how a lack of accessibility to information forces potential tourists to be dependent on others. In this regard, organisations and associations for persons with disability are an important source of information (Calvillo, 2011).

The prevalent interests of researchers in this field have focused on identifying constraints (or barriers) faced by people with disabilities when travelling. There are mainly two approaches in these studies on travel accessibility. Firstly, there is a focus on the analysis of the accessibility of spaces such as cities (Costa & Eniele, 2013), natural spaces (Chikuta *et al.*, 2019; Medina, 2017), religious spaces (Guillén & Ramón, 2015), lodgings (Domínguez *et al.*, 2015; Poria *et al.*, 2011) and activities at the destination (Alixandroae *et al.*, 2014). Secondly, there is a focus on the need to remove barriers at the destination (Calvillo, 2011; IMSERSO, 2002; Kołodziejczak, 2019; UNWTO, 2014), specifically external barriers (architectural impediments, customer service and a lack of specific training or communication skills).

This chapter focuses on the pre-trip stage, in which access to information about the destination and accessibility of resources is a key element in making a decision to travel. Planning and organising a trip is, most of the time, a long and complicated process, especially in the stage of pre-travel, in which information is a key element in making travel decisions. This process is even harder for functionally diverse visitors as they must overcome extra barriers at this stage, such as a lack of information on the destination's accessibility. The aim of this chapter is to analyse the content criteria for providing information about accessibility on online platforms. In accordance with the United Nations World Tourism Organization (UNWTO), the criteria considered were that the information needs to be credible, visible, up-to-date, complete, adequate and universally accessible (UNWTO, 2014).

Buholz (as cited in Kołodziejczak, 2019) defines tourist information as a system comprising an organised set of data for the organisers and consumers of tourist services. According to this definition, when people are planning a trip they expect to receive practical and updated information about the destination; in the case of people with disabilities, they also expect to have accurate information about accessibility to resources at the destination. According to the Spanish Society for the Management of Tourist Innovations and Technologies (SEGITTUR, 2015), content plays a key role in the process of decision-making for people with disabilities.

One of the travelling constraints for people with disabilities is available information about accessibility. This includes information about types of disability, tourist information, tourism development and the convenience of the location (Burnett & Baker, 2001). Many authors have pointed out the important role of information about accessibility, the lack of it and even its complete absence (Buhalis & Amaranggana, 2014; Darcy & Buhalis, 2011; Kołodziejczak, 2019).

According to UNWTO (2015a), there are two main barriers related to information. Firstly, there are barriers regarding the characteristics of communication systems – both hardware and software. Calvillo (2011) and Wang *et al.* (2017) showed how different types of functional diversity face different online barriers, making it possible for a website to be accessible for one group but not for another. Other researchers have studied website accessibility (Fontanet & Mayol, 2011; Martínez *et al.*, 2016), app accessibility (Soares *et al.*, 2017) and the use of social media for information services (Wang *et al.*, 2017). All of these studies are focused on barriers to accessibility arising from the technical design of both hardware and software. Secondly, there are barriers related to information content (Hallett & Kaplan-Weinger, 2010). Studies on barriers related to this issue are diverse (Brilhante & Corrêa, 2015; Gil, 2013; UNWTO, 2015a) and less numerous than those related to web accessibility.

Bowtell (2015) concludes that people with disabilities would travel if they were sure of obtaining reliable, correct, concrete and complete information about the destination. In terms of the content offered by online platforms, it has been noted that there is no international consensus about tourism accessibility standards (Brilhante & Corrêa, 2015; Gil, 2013; UNWTO, 2015a). Furthermore, the information provided is not always properly looked after: some models are based on alerts and traffic signals but offer no details on information, information is not always updated and, sometimes, accessible rooms or spaces are provided but overall inclusion is not considered.

According to the principles established by UNWTO (2015a) regarding the accessibility of any type of information and the principles established in the web content accessibility guidelines (WCAG) 2.0 (Caldwell *et al.*, 2008), we can conclude that online information must facilitate users to find the information they need (visually or through sound), to understand it and, finally, to decide how to use it to be useful for decision-making.

In this chapter, the results of a bibliographic review about key concepts related to accessible tourism are presented. How accessibility and ageing is addressed in United Nations conventions and European and Spanish policies is then described. The main results and conclusions derived from an analysis of tourism apps and official tourism bodies are then presented.

The Evolution of Accessible Tourism: From an Uncertain Past to a Consolidated Future

Accessibility refers to all the characteristics possessed by the environment, product or service, with the necessity for all individuals to be able to move in horizontal and vertical planes, to use objects and to perceive and understand information (CEAPAT, 1996). According to the European concept of accessibility, accessibility is the condition that allows housing, shops, theatres, parks, hotels, hospitals, workplaces and leisure facilities to be reached, entered, exited and used. This applies to buildings (housing, buildings for public use, buildings for leisure use), urban planning (streets), transport (buses, railways, trams and taxis, and air and sea transport), communications and signalling. Accessibility allows individuals to participate in the social and economic activities of their built environment (Domínguez *et al.*, 2019).

Two key aspects should be noted (Darcy & Buhalis, 2011). Firstly, accessibility is useful for everyone, not just those with a disability (Millán, 2010). An accessible environment is one that 'can be used efficiently and safely by the greatest possible number of individuals' (CEAPAT, 2002). The importance of a universal vision of accessibility is justified by the various groups that benefit from it, including the elderly and individuals with disabilities (CEAPAT, 2001; Fernández Alles, 2009a). As a consequence of increases in life expectancy, the ageing population implies a considerable increase in people who want to participate in tourism, many of whom will need special products and services (Garay & Canoves, 2009). Secondly, universal design, or design for all, is a tool for advancing accessibility conditions. Design for all (Fernández Alles, 2009b; Marcos & González, 2002) is an action that involves environments, products and services with the purpose of being able to reach the greatest number of potential clients, taking into account different present and future generations regardless of age, gender, abilities or cultural knowledge.

From a universal accessibility approach, it is worth highlighting the firm steps that have taken place on different fronts and that have allowed considerable progress to be made. Standing out among these are the following.

- Consolidation of the concept of accessibility and design for all (Burnett & Baker, 2001; Millán, 2010).

- The development of subnational regulatory frameworks on barriers and accessibility, and the production of universal accessibility plans.
- Shifting paradigms from removing barriers to promoting accessibility and the development of technical standards (ENAT, 2006).
- Universal accessibility diagnostics and analysis studies, and the publication of manuals to promote accessibility for different types of disabilities and environments (UNWTO, 2014).

Policies in Favour of Functional Diversity: Disability, Ageing and Tourism

Policymaking forerunners to promote and develop accessible tourism date back to 1976 when the Society for Accessible Travel and Hospitality was created in the USA with the aim of promoting and raising awareness about tourism for all, including elderly and other groups with extra needs, such as pregnant women and those travelling with children (Fernández Alles, 2009b). However, it was not fully recognised until the 1990s when the UNTWO considered tourism to be not simply an economic affair, but a fundamental right and a key vehicle for human development UNWTO (2015b).

From the turn of the century, the concept of universal accessibility became widespread and specific terminology emerged in the tourism field: barrier-free tourism, easy-access tourism, disabled tourism, elderly tourism, special needs tourism, accessible tourism, tourism for all and inclusive tourism (Gillovic *et al.*, 2018; Martin *et al.*, 2018; Martínez Cárdenas, 2017; Soret & Barragán, 2015; UNWTO, 2015b). Today, inclusive tourism is understood as 'a type of tourism that involves collaboration between stakeholders, allowing individuals with access requirements, including mobility, vision, hearing, and cognitive dimensions, to be independent with equality and dignity through the provision of tourism environments, products, and services that are universally designed' (Darcy & Buhalis, 2011: 181).

It is worth highlighting, as stated in the World Charter for Sustainable Tourism (IRT, 2015), that the ultimate responsibility for managing accessible tourism should fall upon local communities and destinations.

Two fundamental conclusions are derived from the milestones noted above.

Firstly, despite the remarkable progress made in terms of public policy from the turn of the century onwards, to understand accessibility as a normalised and unnoticed issue is understood as one of the main challenges of policies in favour of functional diversity, including disability, ageing and tourism. As of now, accessibility actions continue to be considered as evident and extraordinary. Furthermore, quality accessibility – understood as that which simplifies everyone's lives, makes products, services, spaces, communications and environments more useful for a greater number of individuals and is offered at low or no additional economic cost through standardised design – goes unnoticed (Rovira-Beleta, 2009).

Secondly, while a single and internationally accepted definition of accessible tourism does not yet exist, evolution of the concept and its application into tourism polices is a growing reality (Fuguet, 2008). Accessible tourism would not be considered a type of tourism at all if there were absolute universal accessibility in all tourism environments, products and services (Hernández-Galán, 2017). Accessibility should be an inherent component of tourism, as is the case with quality, sustainability and so on. Consequently, we should talk about accessible sun and beach tourism or accessible cultural tourism, for example. The challenge lies in building a society that is capable of addressing all its members with inclusive models in all areas and consolidating accessibility in the area of tourism.

Digital Accessibility and Smart Destinations

Accessibility is presently being addressed as a main concept inherent to a new destination model known as 'smart destinations'. One of the key elements in the evolution of tourist destinations to smart tourist destinations is that information and communication technologies (ICTs) are essential tools in the management of any smart destination (Boes *et al.*, 2016; Gretzel *et al.*, 2015; Huertas *et al.*, 2019). Lamsfus *et al.* (2015) note that a destination is not 'smart' because it uses technology, but because it uses technology to understand and facilitate human mobility. In this sense, Cimbaljevic´ *et al.* (2019) consider that ICT are useful in order to exchange information and knowledge.

The smart destination concept also includes accessibility as a key element of a destination's structure, together with technology, innovation and sustainability (Buhalis & Amaranggana, 2014; Burnett & Baker, 2001; European Commission, 2015; Huang *et al.*, 2012; Ivars *et al.*, 2016; UNWTO, 2015c). Most smart destinations have developed accessible services based on smart technology solutions such as safer intersection crossings, automation, robotics, smart wayfinding and navigation systems or integrated payment systems (Mandić, 2019).

Moreover, a smart destination must include smart information systems (Wang *et al.*, 2016). A smart tourism system is a crucial element in a smart destination as it allows different data sources to be analysed, selected according to the destination's needs, and for information to be gathered, processed and analysed, transforming it into knowledge.

A smart tourism system is a strategic tool for destinations to communicate interactively with tourists and enhance the tourism experience (Boes *et al.*, 2016). With the use of smartphones, tourists can obtain information in real time, achieve more satisfactory experiences and co-create experiences (Huertas *et al.*, 2019). In order to stimulate information sharing, destinations should use different technology components as end-user internet service systems, including, for example, destination apps, augmented reality and QR codes.

Smartness has gained significant attention from tourism practitioners, especially in Spain (Lopez de Avila, 2015) and China (Li *et al.*, 2016). In terms of digital accessibility, the Spanish smart tourist destination model (SEGITTUR, 2015: 45) specifically states that 'it must promote the adaptation of all of its digital material, both in terms of its webpages, as well as its online promotional materials, to existing international protocols, including web content accessibility guidelines 2.0 (WCAG 2.0) and the mobile web best practices (MWBP), and mobile web application best practices of the W3C (World Wide Web Consortium)'.

In Spain, this model has been translated into a standard by the Technical Committee on Smart Cities, called AEN/CTN 178, established by the Spanish Association for Standardization and Certification, and took on a practical dimension for destinations with the Smart Tourist Destinations Network being formally established in February 2019 by the Spanish Tourism Secretary of State.

As already noted, following the principles established by UNWTO (2015a) for any type of information and those established in WCAG 2.0 (Caldwell *et al.*, 2008), for online information to be useful for decision-making, it must facilitate users being able to perceive the information (visually or through sound), find the information they require, understand that information, decide how to use it and act as a result.

Therefore, the concept of smart tourism destinations leads to the need for a technological platform on which information about tourism activities can be offered and exchanged instantly (Gajdošík, 2019). Such a platform will also facilitate the dissemination of accessibility information in a suitable way.

Diagnostic Study of Accessibility Information Provided by Tourism Destination Websites

Criteria used to analyse content information

The aim of the research presented in the following was to analyse the content criteria for providing information about accessibility on online platforms. In accordance with UNWTO (2014), the criteria considered to reach that objective were that the information needs to be credible, visible, up-to-date, complete, adequate and universally accessible.

- *Credibility.* This dimension is related to aspects that offer reasonable grounds for being believed, allowing users to assess the degree of credibility of a website. Credibility is associated with website owners, hence the name of the institution or organisation was the initial data gathered.

- *Visibility.* The purpose of this criterion is to determine the visibility of information on the website or app. In other words, to identify whether accessibility information is easily visible at the home page. Information about accessibility must be readily available and anyone should be able to find it without any kind of difficulty. To measure this criterion, we analysed (1) the level of the website map from which information about accessibility is accessed, (2) the format of content information (pdfs, pictograms, icons, links, etc) and (3) the existence on the home page of a search engine that enables a user to find information about accessibility.
- *Updatedness.* Keeping website information up-to-date is a key issue. Therefore, references to the date the resource was last revised were used as an indicator.
- *Correctness and completeness.* Accuracy of information is the most obvious criterion for the quality of content: users have the right to expect that websites and apps will provide accurate information. This was measured by assessing the presence of information relevant to users' needs.

The indicators used for the assessment were identified from an analysis of previous studies (Buenadicha *et al.*, 2001; Huizingh, 2000; Kozak *et al.*, 2005; Miranda & Bañegil, 2004; Miranda *et al.*, 2006, 2015), while also taking into consideration the experience of the research team. The indicators used were:

- general information;
- languages in which information is offered;
- use of easy language;
- the types of disability on which information is included;
- how information is offered;
- if it allows for transportation, lodging, routes or itineraries to be planned, or tickets to, for example, shows or museums to be purchased.

All European countries were considered in the sample. The official tourism websites of both countries and capital cities were was taken as sources of information. A Boolean search on Google combining the words 'country' OR 'capital' AND 'accessible' OR 'for all' OR 'barrier-free' as Boolean operators was used to define searches. In addition, as mobile devices are commonly used in the stage of pre-travel, travel apps for the most popular mobile operation systems (Android and iOS) were also included in the study, using the search words 'accessible tourism' or 'tourism for all'.

Analysis of websites and apps

In total, 147 websites and apps from European countries and international organisations were identified. The main results of the analysis, carried out in December 2018, are now presented.

Regarding credibility, it is worth noting that most of the analysed websites and apps were managed by public administration entities in charge of tourism for each country or city or by international entities such as the European Network for Accessible Tourism or Europe for All. According to Kakol *et al.* (2017), the most important factor, and at the same time the easiest one to analyse regarding the credibility of a website is, precisely, that it is an official site.

Moreover, there are a large number of sites and apps where content is directly uploaded by diverse ability users who relate their experiences (e.g. Wheelchair Travel, Sage Traveling and Equalitas Vitae). These type of platforms are responding to the lack of reliable information on tourism destination accessibility on the official sites and the need to count on the actual experiences of users in order to ensure the credibility of information provided (Herrera-Viedma & Peis, 2003; Herrera-Viedma *et al.*, 2006).

Regarding content visibility, only two countries (Czech Republic and Hungary) did not offer specific information on accessibility on their official tourism websites or apps.

With regard to updatedness, the indicator 'date resource last revised' is a good practice that, unfortunately, is not as widely employed as it should be.

Concerning content adequacy, in most cases, information was presented in descriptive form (through text) on the destination's tourism website or app, sometimes accompanied by pictograms, icons or videos to facilitate interpretation of the information. In some cases, links were used to refer to other websites for accommodation, restaurants, resources, places of interest or major modes of transport. Of the websites analysed, 43% referred to information in a pdf file. However, these files are not updated, offer static information and are not easy to understand by all people with disabilities. For instance, there is often a lack of contrast in pdfs, so partially sighted people find them hard to follow, or they do not follow established easy reading policies, so people with cognitive disabilities cannot use them either. Only five of the websites analysed offered the possibility of interacting with the information provided.

In terms of the languages in which the information was offered, English was found to be the predominant language (97.9%), offered together with the platform's native language. Although general information about the destinations was offered in other languages, the same was not true of information specific to accessibility.

In terms of the disabilities that the platforms considered, all of them with information on accessibility discussed physical disability. The majority also made reference to hearing and visual impairment. However, only five of the platforms analysed took cognitive disability into consideration and included easy-to-understand language. Only seven of the platforms included accessibility for individuals with mental health problems. Persons with disability are a heterogeneous group that must be considered, with all of their needs addressed.

Only 14 of the platforms (9.5% of the total analysed) included testimonials from persons with a disability. In a few cases, ratings and opinions from comparison sites such as Tripadvisor and social media were included. However, these opinions were general comments that did not include accessibility as a search filter.

We thus concluded that the information offered by most platforms analysed is not adequately designed to facilitate the browsing of its contents or the finding and reading of information. Functionally, people with disabilities not only face a lack of information, but also disinformation, understood as information that is contradictory, unreliable, out-of-date or inaccurate. This uncertainty is overcome by finding other trustworthy sources such as associations and organisations related to the individual's disability.

Conclusions

Most web evaluation models and methodologies (Aladwani & Palvia, 2002; Dhyani *et al.*, 2002; Kirakowski & Cierlik, 1998; López *et al.*, 2010; Negash *et al.*, 2003; Olsina & Rossi, 2002; Olsina *et al.*, 2006; Powell *et al.*, 1998; Sellito & Burgess, 2005) tend to be more objective than subjective, quantitative rather than qualitative and do not take into account user perception (Dhyani *et al.*, 2002). Despite acknowledging its complexity, any website quality evaluation methodology should include the perception of users. Furthermore, the aspects of quality to be considered and their relative importance vary according to the application domain and the goals to be achieved by the user (Moreno *et al.*, 2009). In the present study, users were considered to be people with diverse abilities going online with the aim of obtaining information about the accessibility of a destination.

The quality of websites can only be evaluated if user perceptions are gathered, which is normally not an easy task. It is necessary to understand whether websites comply with users' needs and expectations and the user-perceived quality of accessibility information on tourism websites becomes especially relevant given the potential implications on destination management.

In order to make decisions about trips, users must be able to ensure that the travel chosen offers the support, assistance and infrastructure they need from the beginning to the end of the trip. There must be accurate, detailed and up-to-date information to suits their needs. However, the results of this study indicate that the information offered by most tourism websites analysed does not provide the reliable information necessary to plan an entire itinerary.

It is important that visitors are backed by a system that guarantees reliable information on whether places and activities are accessible, adapted, inclusive for all and can be enjoyed by everyone – a system that also checks the information meets real-life needs and ensures it is accessible, up-to-date, precise, correct and accurate. Ultimately, so that all visitors and users can make decisions about their travels based on that information in full confidence and without false expectations.

The involvement of all people exchanging knowledge and skills is essential to for co-creation and the success of smart tourism destinations. The information society can offer a collection of new tools to access a wealth of information. The future of information about accessibility points to crowdsourcing (Wang *et al.*, 2017), which would enable travellers with disabilities to contribute and discover accurate information on the accessibility of destinations. The innovative aspect of crowdsourcing lies in its community-based aspect: organisations, administrators, host communities and tourists – with and without disabilities – collecting and sharing local information to contribute to the updating, relevance and accuracy of easily available accessibility information. In this way, people with disabilities will be empowered by producing and consuming information at the same time, creating a data collection that future visitors and local residents with disabilities will benefit from. Consumers become 'prosumers' because they have the knowledge, desire the ability to co-create information with tourism service providers.

References

Aladwani, A. and Palvia, P. (2002) Developing and validating an instrument for measuring user-perceived web quality. *Information & Management* 39 (6), 467–476.

Alixandroae, I., Dobre, L., Comănescu, L. and Nedelea, A. (2014) Evaluating the landscape accessibility for tourism activities in Postăvaru mountains. *Studia Universitatis Babeş-Bolyai: Geographia* LIX (2), 157–166.

Boes, K., Buhalis, D. and Inversini, A. (2016) Smart tourism destinations: Ecosystems for tourism destination competitiveness. *International Journal of Tourism Cities* 2 (2), 108–124.

Bowtell, J. (2015) Assessing the value and market attractiveness of the accessible tourism industry in Europe: A focus on major travel and leisure companies. *Journal of Tourism Futures* 1 (3), 203–222.

Brilhante, M.N. and Corrêa, C. (2015) Análise comparativa de guias turísticos em formato de aplicativo: Lonely Planet e mTrip. *Revista Turismo – Visão e Ação – Eletrônica* 17 (2), 354–386.

Buenadicha, M., Chamorro, A., Miranda, F.J. and González, O.R. (2001) A new web assessment index: Spanish universities analysis. *Internet Research* 11 (3), 226–234, doi:10.1108/10662240110396469.

Buhalis, D. and Amaranggana, A. (2014) Smart tourism destinations. In Z. Xiang and I. Tussyadiah (eds) *Information and Communication Technologies in Tourism* (pp. 553–564). Cham: Springer International. doi: 10.1007/978-3-319-03973-2_40

Burnett, J.J. and Baker, H. (2001) Assessing the travel-related behaviors of the mobility-disabled consumer. *Journal of Travel Research* 40 (1), 4–11.

Caldwell, B., Cooper, M., Guarino Reid, L. and Vanderheiden, G. (eds) (2008) Web content accessibility guidelines (WCAG) 2.0. See https://www.w3.org/TR/WCAG20/ (accessed January 2021).

Calvillo, M. (2011) Orientaciones para el desarrollo y rentabilidad de productos turístico accesibles a partir del estudio de las necesidades y hábitos turísticos de las personas con discapacidad. In J.L. Jiménez and P. De la Fuente (eds) *Turismo y desarrollo económico: IV jornadas de investigación en turismo* (pp. 537–565). Sevilla: Editorial Digital Altres.

CEAPAT (Centro Estatal de Autonomía, Personas y Ayudas Técnicas) (1996) *Concepto Europeo de Accesibilidad*. Madrid: Ministerio Trabajo y Asuntos Sociales and CEAPAT.

CEAPAT (2001) *Resolución del Consejo Europeo sobre el Diseño Universal*. Madrid: Ministerio Trabajo y Asuntos Sociales.

CEAPAT (2002) *Criterios de Diseño en Productos para Mayores*. Madrid: Ministerio Trabajo y Asuntos Sociales.

Chikuta, O., Du Plessis, E. and Saayman, M. (2019) Accessibility expectations of tourists with disabilities in National Parks. *Tourism Planning & Development* 16 (1), 75–92.

Cimbaljević, M., Stankov, U. and Pavluković, V. (2019) Going beyond the traditional destination competitiveness – reflections on a smart destination in the current research *Current Issues in Tourism* 22 (20), 2472–2477.

Cole, S., Zhang, Y., Wang, W. and Hu, C. (2019) The influence of accessibility and motivation on leisure travel participation of people with disabilities. *Journal of Travel & Tourism Marketing* 36 (1), 119–130.

Costa, D. and Eniele, K. (2013) Turismo accesible en la estructura urbana de las ciudades turísticas: El caso de Santa Cruz, RN – Brasil. *Estudios y Perspectivas en Turismo* 22 (6), 1045–1073.

Darcy, S. and Buhalis, D. (2011) Conceptualising disability. In D. Buhalis and S. Darcy (eds) *Accessible Tourism: Concepts and Issues* (pp. 21–42). Bristol: Channel View Publications.

Dhyani, D., Ng, W.K. and Bhowmick, S. (2002) A survey of web metrics. *ACM Computing Surveys* 34 (4), 469–503.

Domínguez, T., Fraiz, J.A. and Alén, M.E. (2015) Discapacidad y alojamientos turísticos en España. *Pasos: Revista de Turismo y Patrimonio Cultural* 13 (4), 771–787.

Domínguez, T., Alén, M.E. and Darcy, S. (2019) Accessible tourism online resources: A Northern European perspective Scandinavian. *Journal of Hospitality and Tourism* 19 (2), 140–156, doi:10.1080/15022250.2018.14.

European Commission (2015) Smart cities. See http://ec.europa.eu/eip/smartcities/ (accessed March 2020).

ENAT (European Network of Accessibility Tourism) (2006) *Trabajando juntos para hacer el Turismo en Europa Accesible a Todos*. Madrid: Fundación ONCE and ENAT.

Fernández Alles, M.T. (2009a) El diseño universal: Concepto y certificación. *Boletin del Real Patronato* 75, 4–11.

Fernández Alles, M.T. (2009b) Turismo accesible: Importancia de la accesibilidad para el sector turístico. *Entelequia. Revista Interdisciplinar* 9, 2–16.

Fontanet, G. and Mayol, J. (2011) Importancia y situación actual de la accesibilidad web para el turismo accesible. *Pasos: Revista de Turismo y Patrimonio Cultural* 9 (2), 317–326.

Fuguet, T. (2008) Europa demanda más accesibilidad. *Editur* 7, 10–15.

Gajdošík T. (2019) Towards a conceptual model of intelligent information system for smart tourism destinations. In R. Silhavy (ed.) *Software Engineering and Algorithms in Intelligent Systems. Advances in Intelligent Systems and Computing* (Vol. 763, 66–74). Cham: Springer. https://doi.org/10.1007/978-3-319-91186-1_8.

Garay, L. and Canoves, G. (2009) El desarrollo turístico en Cataluña en los dos últimos siglos: Una perspectiva transversal. *Documents d'Analisi Geografica* 53, 29–46.

Gil, S. (2013) *Cómo hacer "Apps" accesibles*. Madrid: CEAPAT.

Gillovic, B., McIntosh, A., Darcy, S. and Cockburn-Wootten, C. (2018) Enabling the language of accessible tourism. *Journal of Sustainable Tourism* 26 (4), 615–630.

Gretzel, U., Koo, C., Sigala, M. and Xiang, Z. (2015) Special issue on smart tourism: Convergence of information technologies, experiences, and theories. *Electronic Markets* 25 (3), 175–177.

Guillén, L. and Ramón, A. (2015) Valoración de la infraestructura de los edificios religiosos para el turismo accesible en Villahermosa, Tabasco, México. *Pasos: Revista de Turismo y Patrimonio Cultural* 13 (3), 491–508.

Hallett, R. and Kaplan-Weinger, J. (2010) *Official Tourism Websites: A Discourse Analysis Perspective*. Bristol: Channel View Publications.

Hernández-Galán, J. (2017) *Observatorio de Accesibilidad Universal del Turismo en España*, Madrid: Fundación ONCE and Via Libre.

Herrera-Viedma, E. and Peis, E. (2003) Evaluating the informative quality of documents in sgml-format using fuzzy linguistic techniques based on computing with words. *Information Processing & Management* 39 (2), 195–213.

Herrera-Viedma, E., Pasi, G., Lopez-Herrera, A. and Porcel, C. (2006) Evaluating the information quality of web sites: A methodology based on fuzzy computing with words. *Journal of the American Society for Information Science and Technology* 57 (4), 538–549.

Huang, X.K., Yuan, J.Z. and Shi, M.Y. (2012) Condition and key issues analysis on the smarter tourism construction in China. In F.L. Wang, J. Lei, R.W.H. Lau and J. Zhang (eds) *Multimedia and Signal Processing* (pp. 444–450). Berlin: Springer. https://doi.org/10.1007/978-3-642-35286-7_56.

Huertas, A., Moreno, A. and Hamy, T. (2019) Which destination is smarter? Application of the (SA)6 framework to establish a ranking of smart tourist destinations. *International Journal of Information Systems and Tourism* 4 (1), 19–28.

Huizingh, E. (2000) The content and design of web sites: An empirical study. *Information & Management* 37, 123–134, doi:10.1016/S0378-7206(99)00044-0.

Ivars, J.A., Solsona, F.J. and Giner, D. (2016) Gestión turística y tecnologías de la información y la comunicación (TIC): El nuevo enfoque de los destinos inteligentes. *Documents D'Anàlisi Geogràfica* 62 (2), 327–346.

IMSERSO (Instituto de Mayores y Servicios Sociales) (2002) *Libro Verde de la accesibilidad en España*. Madrid: IMSERSO.

IRT (Institute of Responsible Tourism) (2015) World charter for sustainable tourism. See http://sustainabletourismcharter2015.com/the-world-charter-for-sustainable-tourism/ (accessed March 2020).

Kakol, M., Nielek, R. and Wierzbicki, A. (2017) Understanding and predicting web content credibility using the content credibility corpus. *Information Processing & Management* 53 (5), 1043–1061.

Kirakowski, J. and Cierlik, B. (1998) Measuring the usability of web sites. *Proceedings of Human Factors and Ergonomics Society 42nd Annual Meeting, Santa Monica, CA, USA*. https://doi.org/10.1177/154193129804200405

Kołodziejczak, A. (2019) Information as a factor of the development of accessible tourism for people with disabilities. *Quaestiones Geographicae* 38 (2), 67–73.

Kozak, M., Bigné, E. and Andreu, L. (2005) Web-based national tourism promotion in the Mediterranean area. *Tourism Review* 60 (1), 6–11, doi:10.1108/eb058447.

Lamsfus, C., Martin, D., Alzua-Sorzabal, A. and Torres-Manzanera, E. (2015) Smart tourism destinations: An extended conception of smart cities focusing on human mobility. In L. Tussyadiah and A. Inversini (eds) *Information and Communication Technologies in Tourism* (pp. 363–375). Cham: Springer.

Li, Y., Hu, C., Huang, C. and Duan, L. (2016) The concept of smart tourism in the context of tourism information services. *Tourism Management* 58, 293–300, doi:10.1016/j.tourman.2016.03.014.

López, J., Chica, J., Arcila, M., Azzariohi, A. and Soto, A. (2010) Modelo de análisis de las páginas web turísticas (diputaciones y ayuntamientos de las capitales de provincia de Andalucía). *Historia Actual Online* 22 (1), 185–200.

Lopez de Avila, A. (2015) Smart destinations: XXI century tourism. Presented at *ENTER2015 Conference on Information and Communication Technologies in Tourism, Lugano, Switzerland* (No published).

Mandić, A. (2019) Progress on the role of ICTs in establishing destination appeal: Implications for smart tourism destination development. *Journal of Hospitality & Tourism Technology* 10 (4), 791–814.

Marcos, D. and González, D.J. (2002) *Turismo Accesible Hacia un Turismo para Todos*. Madrid: Mazars.

Martin, M.C., Luque, E. and de la Fuente, Y. (2018) Turismo inclusivo para todas las personas. Una apuesta por la diversidad. *REID Monográfico* 3, 81–96.

Martínez, A.M., Roselló, E. and Cifuentes, R. (2016) *Calidad de Websites Turísticas Oficiales de la Costa Mediterránea Española (2004–2015)* [Doctoral Dissertation, Universidad de Murcia]. Universidad de Murcia. Murcia, España.

Martínez Cárdenas, R. (2017) Turismo accesible en México. *Revista Científica sobre Accesibilidad Universal, La Ciudad Accesible* 14 (9), 23–33.

Medina, M. (2017) Propuesta de desarrollo del turismo accesible en la reserva de biósfera Isla de Ometepe (Nicaragua). *Pasos: Revista de Turismo y Patrimonio Cultural* 15 (4), 913–924.

Millán, M. (2010) Turismo accesible/turismo para todos, un derecho ante la discapacidad. *Gran Tour: Revista de Investigaciones Turísticas* 2, 101–126.

Miranda, F.J. and Bañegil, T.M. (2004) Quantitative evaluation of commercial web sites: An empirical study of Spanish firms. *International Journal of Information Management* 24 (4), 313–328, doi:10.1016/j.ijinfomgt.2004.04.009.

Miranda, F.J., Rubio, S. and Chamorro, A. (2006) El uso de Internet por los Operadores Logísticos. *Cuadernos de Estudios Empresariales* 16, 99–114.

Miranda, F.J., Rubio, S. and Chamorro, A. (2015) The web as a marketing tool in the Spanish foodservice industry: Evaluating the websites of Spain's top restaurants. *Journal of Foodservice Business Research* 18 (2), 146–162, doi:10.1080/15378020.2015.1029386.

Moreno, JM., Morales del Castillo, J.M., Porcel, C. and Herrera-Viedma, E. (2009) A quality evaluation methodology for health-related websites based on a 2-tuple fuzzy linguistic approach. *Soft Computing* 14, 887–897.

Negash, S., Ryan T. and Igbaria, M. (2003) Quality and effectiveness in web-based customer support systems. *Information & Management* 40, 757–768.

Olsina, L. and Rossi, G. (2002) Measuring web application quality with webQEM. *IEEE Multimedia* 9 (4), 20–29.

Olsina, L., Covella, G. and Rossi, G. (2006) Web quality. In E. Mendes and N. Mosley (eds) *Web Engineering* (pp. 109–142). Berlin: Springer.

Poria, Y., Reichel, A. and Brandt, Y. (2011) Dimensions of hotel experience of people with disabilities: An exploratory study. *International Journal of Contemporary Hospitality Management* 23 (5), 571–591.

Powell, T., Jones, D. and Cutts, D. (1998) *Web Site Engineering: Beyond Web Page Design*. Englewood Cliffs, NJ: Prentice Hall.

Rovira-Beleta, E. (2009) *Personas, Dependencia, Calidad de vida y Nuevas Tecnologías*. Barcelona: Ed Hacer.

SEGITTUR (2015) *Libro Blanco de los Destinos Turísticos Inteligentes*. Madrid: SEGITTUR. https://www.segittur.es/wp-content/uploads/2019/11/Libro-Blanco-Destinos-Tursticos-Inteligentes.pdf

Sellito, C. and Burgess, S. (2005) Towards a weighted average framework for evaluating the quality of web-located health information. *Journal of Information Science* 31 (4), 260–272.

Soares, J., Gabriel, L. and Sánchez, M. (2017) Un análisis de la app turística Tenerife accesible. *Podium* 6 (1), 109–123.

Soret, P. and Barragán, A. (2015) Turismo accesible y legislación. *Estudios Turísticos* 203–204, 75–85.

Soriano, L.I. (2017) El turismo accesible como respuesta a una oportunidad de mercado en El Salvador. *Teoría y Praxis* 30, 85–99.

UNWTO (United Nations World Tourism Organization) (2014) *Manual Sobre Turismo Accesible Para Todos: Principios, Herramientas y Buenas Prácticas*. Madrid: UNWTO.

UNWTO (2015a) *Recomendaciones de la OMT Sobre Accesibilidad de la Información Turística (A/RES/669(XXI)), adoptadas por la Asamblea General de la OMT, el día 17 de septiembre de 2015, en Medellín, Colombia*. Madrid: UNWTO.

UNWTO (2015b) *Turismo Para Todos: Promover la Accesibilidad Universal*. Madrid: UNWTO.

UNWTO (2015c) *UNWTO Committee on Tourism and Competitiveness*. Madrid: UNWTO.

Wang, W., Chih-Hung, Y., Xiong, H. and Wan-Ju, L. (2017) Use of social media in uncovering information services for people with disabilities in China. *International Review of Research in Open and Distance Learning* 18 (1), 65–83.

Wang, X., Li, X.R., Zhen, F. and Zhang, J. (2016) How smart is your tourist attraction? Measuring tourist preferences of smart tourism attractions via a FCEM-AHP and IPA approach. *Tourism Management* 54, 309–320.

6 Current Perspectives on Social Inclusion in Tourism in Finnish Lapland

Sari Nisula, Marlene Kohllechner-Autto and Krista Skantz

Introduction

The importance of sustainability is growing rapidly in Europe. Surveys and trends show that the sustainable way of doing business is an increasingly important factor for customers. In the autumn of 2018, Business Finland organised a tour operator survey and found that tour operators were willing to increase the supply of sustainably produced tourism experiences. Climate change is on customers' minds, with growing demand for climate-friendly products, especially those with a low carbon footprint (Business Finland, 2019).

The increasing importance of sustainability in business is borne out in other surveys and trends. A survey conducted by Reiseanalyse in 2014 (quoted in Business Finland (2019)) showed that travellers who value sustainability travel more often, stay longer in the destination, spend more on their journey and a third are willing to pay more for sustainable services. In addition, tourists are looking for places where the air is clean, the senses are perceptible and there is a chance to meet the locals (Business Finland, 2019). The principles of sustainable development should thus be reflected in all marketing communications concerning Finland. Sustainability is also closely related to highlighting the competitive advantage of Finland's main attraction and 'pureness' (Business Finland, 2019).

The United Nations World Tourism Organization (UNWTO) defines sustainable development in tourism as:

> Tourism that takes full account of its current and future economic, social and environmental impacts, addressing the needs of visitors, the industry, the environment and host communities. (UNWTO, 2020)

The World Travel and Tourism Council (WTTC) uses the United Nations' 17 Sustainable Development Goals (SDGs) released in 2015 to describe the sustainability aspect of tourism. The SDGs are further divided in 169 more specific targets. Although tourism and hospitality explicitly appear in just three of these targets, the WTTC sees tourism and the travel industry as having an important role in the realisation of all 17 SDGs (WTTC, 2020).

As indicated in the above UNWTO definition, sustainable tourism includes also a social dimension. One definition for social sustainability is:

> Development that is compatible with harmonious evolution of civil society, fostering an environment conducive to the compatible cohabitation of culturally and socially diverse groups while at the same time encouraging social integration, with the improvements in the quality of life for all segments of the population. (Polese & Stren, 2000: 15–16)

Taking social sustainability and social inclusion into account when viewing balanced economic growth and social economy regionally, employment and education are some of the key roles (Kohl, 2006). As Dreher *et al.* (2013: 286) point out '[s]ocial inclusion involves a movement towards creating equal opportunities for all citizens, enabling opportunities that can guarantee access to the different needs for the well-being and quality of life; among these elements is access to work'.

Numerous examples of tourism enterprises and social inclusion can be found in several European regions. Examples of enterprises that specifically highlight social inclusion in their communications are Universo Santi restaurant, Brownies and Downies cafe and Magdas Hotel, 2019.

Universo Santi, located in the Spanish city of Jerez, employs people with disabilities in its kitchens. The restaurant aims to improve the quality of life and living conditions of people with disabilities. The message of social inclusion is clearly communicated in its posts on social media and on the restaurant's webpages. The website of Universo Santi states: '… we need to promote the full labour and social integration of people with any type of disability, according to the United Nations, which defends the achievement of full labour integration, which, in the case of people with disabilities, finds greater difficulties' (Universo Santi, 2019, translation).

Another example of a catering enterprise focusing on social inclusion is in the Dutch city of Veghel. Its name – Brownies and Downies cafe – gives an indication of the inclusive nature of the enterprise. The concept of the cafe is to employ people with Down syndrome and other disabilities in a hospitality setting. By 2018, the concept had grown to 53 locations in three locations, with 85 entrepreneurs employing 1183 people with special needs. The growth strategy was implemented through franchises (Brownies and Downies, 2019a). Aside from employment, social inclusion is also visible in the enterprise's supply chain. Tea and fresh produce, for instance, are supplied by partners emphasising social inclusion through the employment of people with disabilities and the reintegration of the unemployed into the labour market (Brownies and Downies, 2019b). The franchise uses a number of different media channels in their communications: apart from social media channels such as Instagram and Facebook, the franchise also runs a blog and shares success stories on its website (Brownies and Downies, 2019a).

In like manner, Magdas Hotel in Vienna is implementing social inclusion in its operations by employing people from refugee backgrounds. As stated in the hotel's concept, refugees still face challenges finding employment in the open labour market: '[for] people with a refugee background, it is still difficult to find work in Austria. Initial lack of German language skills, the resentment of many employers, as well as the circumstance that refugees are only allowed to accept work after receiving a positive response to an asylum application (which can often take months or years) make integration difficult' (Magdas Hotel, 2019). This socially inclusive approach is not only used to facilitate the integration of refugees into the labour market: it also holds a special position in the Viennese hotel market. The hotel made use of its socially inclusive approach to raise money via crowdfunding for its conversion from a former retirement home to its current form (Magdas Hotel, 2019).

What these three examples have in common is that they openly and actively share their stories and concepts of social inclusion. Social inclusion is the cornerstone of the enterprises' operations and is thusly communicated. The slogan of the Viennese hotel sums up the purpose and intent of all three presented examples best, as 'solving social problems with economic means'.

The purpose of this chapter is to investigate whether tourism may be used as a tool to promote social inclusion in Finnish Lapland and to shine a spotlight on examples that do so already. Social inclusion is explored through the examples of three different organisations that act both in tourism and in social inclusion. Two of the examples are Lappish tourism businesses, while the third is an education institution offering vocational special needs education for young people and adults.

Methodology

The material for this chapter was gathered in the form of a review of the literature on sustainable tourism, social economy and social inclusion, along with three focused interviews. The interviews were conducted with representatives of businesses or organisations that are already touching on the subject of sustainability in their everyday operations in different ways. Social inclusion in tourism is still fairly new in Lapland and thus the number of enterprises actively and openly employing social inclusion practices is limited. More than three enterprises were contacted for interviews, but not all were willing or able (due to internal policies) to participate in the study. The organisations for interview were chosen based on the following criteria: the enterprise is linked to tourism, social inclusion is part of the enterprise's sustainable operational practices, it is located in Lapland and is a viable business operation.

The interviews were recorded and transcribed, and the collected material was subsequently analysed using content analysis. Consequently, a process of finding themes and similarities in the interviews was undertaken. For example, one theme that occurred in all the interviews was the lack of workforce in tourism, as we anticipated from the statistics collected by regional and national bodies (see the section 'Lapland as Place of Residence and Tourism Destination, Rapid Tourism Development and High Unemployment Rate').

This chapter is not representative of the situation of social inclusion in tourism in Lapland. The three conducted interviews are not sufficient to provide a comprehensive review on the topic of inclusion in tourism in respect to employment in Lapland; however, neither is this the intention of this chapter. The interviews, nevertheless, shed light on the topic and provide a viewpoint from which to glance the future.

For the sake of anonymity, the enterprises and organisations interviewed are not named. They are, however, identified as two tourism enterprises and one vocational education and training (VET) provider, each located in a different part of Lapland. The tourism enterprises provide accommodation and activity services in a rural setting, while the VET provider specialises in students with learning disabilities.

Social Inclusion and the Social Economy

Inclusion as a phenomenon and terminology in society has its roots in social sciences and social policy. Social inclusion has long played an important part in European social policies, which are a crucial part of the fabric of society in welfare states such as Finland. The definition of social inclusion is not unambiguous: it is complex, multidimensional and context-dependent. The term inclusion is often set against the term exclusion, and both usually manifest themselves in debates around citizenship, solidarity and cohesion of society as well as social inequality. Politically, social inclusion has been seen to be important in a societal context, and more concretely in the context of employment and health policies (Leemann *et al.*, 2015).

Early discussions about social inclusion were initiated by the French sociologist Emil Durkheim, who discussed inclusion and exclusion in the context of societal solidarity. Later, the definition was shaped in social policy debates to prevent the exclusion of groups in danger of marginalisation in society. In the late 20th century, the influence of the social inclusion debate spread widely in Europe and, by the millennium, many national and international contracts, projects and programmes were in place with the aim of involving those population groups living in the margins of society (Leemann *et al.*, 2015).

In Finland, social inclusion is an important part of national well-being and development programmes and has been included in government programmes. It is seen as a tool to tackle poverty and exclusion, and enhance justice and equality in society. In Finnish literature, social inclusion is often perceived as an individual, experiential and emotional phenomenon. The National Institute for Health and Welfare (NIHW) describes social inclusion as follows: 'inclusion in society means everyone's opportunity to health, education, work, livelihood, housing and social relationships' (NIHW, 2018). In addition, it is described as a changing and dynamic process that tackles poverty and exclusion, enables participation society and offers possibilities and resources. It enhances skills and capabilities and guarantees a dignified life (NIHW, 2018).

As already noted, social inclusion is multidimensional phenomenon. The NIHW (2019) uses the division of *having*, *acting* and *belonging* introduced by Helka Raivio and Jarno Karjalainen (2013) in describing what constitutes social inclusion. Raivio and Karjalainen (2013) applied the division of *having*, *loving* and *being* for well-being by sociologist Erik Allardt in their thinking. Allardt described these categorisations consisting of further components (Allardt, 1980: 50). For example, having consists of components such as income, housing, employment, education and health. Loving consists of connection to the local community, family connection and friendships. Being comes together as status, indispensability, political resources and interesting free-time activities. Allardt also mentions that an interview study conducted in Nordic countries indicated very weak – if even existing – connections between the aspects of having, loving and being (Allardt, 1980: 10). Measurable well-being does not inevitably correlate with the subjective experience of well-being.

The dimensions adapted by Raivio and Karjalainen (2013) condense the needs of human and humane existence. *Having* means not only having the possibility of an adequate livelihood, but also well-being and safety. *Acting* means being an active and recognised actor or subject in, for example, matters concerning oneself, being a part of decision-making and having a say in things. *Belonging* is being a part of communities, being accepted and being able to participate. Thus, the having of Raivio and Karjalainen (2013) includes very similar things to Allardt's having, belonging could be interpreted to be closer to loving, and acting to being. Generally, both categorisations illustrate the subject of this text and could be used as a backdrop for the observations of our own interviews. The categorisations of Raivio and Karjalainen (2013) are adapted to reflect the dimensions of social inclusion, which is a more political term than Allardt's view on well-being, and thus puts more emphasis on the capabilities of the individual of being an active citizen. Opposite of the term inclusion is exclusion, and these dimensions also help in describing the features of exclusion. In exclusion, these dimensions are not realised. Exclusion entails economic and health problems as well as insecurity, being pushed aside and not having a part in communities, as well as alienation and objectification.

To implement all the dimensions of social inclusion, policies on a national level need to be drawn up in co-creation with the governing forces and civil society. The EU drives active labour market policies and inclusive work policies, which enhance equal possibilities for citizens to enter and participate in the labour market. Their implementation requires enhancing social employment and work rehabilitation. Hence, it is crucial that services are offered to and within reach of people at risk of exclusion (Leemann et al., 2015).

Concentrating Forces in Social Inclusion: Social Economy Enterprises

In addition to what could be called traditional or mainstream enterprises, social inclusion can be advanced through different forms of social economy enterprises. In the EU, social economy accounts for the employment of more than 11 million people. Social economy enterprises have different social forms and objectives, encompassing a multitude of sectors, such as agriculture and banking. The main objective of traditional social economy enterprises is not to generate profit for their shareholders, but to serve their members. They are managed on the principle of one vote per person and their members act in accordance with solidarity and mutuality principles. Social economy enterprises fall into four categories: social enterprises, associations and foundations, mutual societies and cooperatives (European Commission, 2019).

In Finland, the concept of social enterprise is still largely a rarely used one. This is due in part to Finland being a Nordic social welfare state, in which either the state or municipalities take care of services provided in other countries by social enterprises. However, social entrepreneurship is increasingly being advanced via, for example, a certificate for social enterprise managed by the Association for Finnish Work (AFW) and the networking efforts of Arvo-liitto (https://arvoliitto.fi/). These two organisations define social entrepreneurship in a similar fashion: enterprises that direct most of their profit into solving a societal problem. These problems might be related to the environment, employment, housing health or the problems faced by elderly people, for example. Company form is not limited. Certification by the AFW aims to raise awareness of social enterprises and to help people find companies that pursue improving society. The social enterprise certificate is not the only certification managed by the AFW. The AFW also manages the Key Flag symbol, which communicates the Finnish origin of products for consumers – it is very well-known symbol in Finland, with almost everyone recognising it and knowing its meaning (AFW, 2019).

Although social enterprises do not (yet) play as significant a role in Finland as they do in other countries, cooperatives have a long tradition in the country. Cooperatives – as defined by the International Cooperative Alliance (ICA, 2016) – are independent associations whose members have mutually agreed on joining forces to gain economic, social and cultural advantage through a democratically led enterprise. While cooperatives are not automatically also social enterprises, their principles and values, such as social responsibility, equality and solidarity, are closely linked to social enterprises and social justice in general (Harju-Myllyaho et al., 2017).

EU-wide initiatives to promote social entrepreneurship can also be found at project level. The Social Entrepreneurship in Sparsely Populated Areas (SOCENT SPAs) project, funded by Interreg Europe, is one of many projects active in the sphere of social economy and social entrepreneurship. The project aims at developing the participating regions' competitive edge by promoting social entrepreneurship in sparsely populated areas. The project's partnership consists of public and private entities from Spain, Germany, Slovakia and Finland. During its five-year duration, the action will foster interregional cooperation with a view to improving the effectiveness of respective regional policies by actively supporting the visibility, incubation and acceleration of social enterprises in sparsely populated areas as a driver for regional competitiveness and inclusive growth (SOCENT SPAs, 2017).

The current situation of social enterprises in Lapland was investigated in the frame of the SOCENT SPAs project. The investigation revealed that the concept of social enterprise is still a fairly unknown one in Lapland and only a handful of social enterprises currently exist. However, many organisations in Lapland and Finland possess the general characteristics of social enterprises, without labelling themselves as such or being even aware of such a characterisation (Harju-Myllyaho et al., 2017).

Viewing social inclusion in the context of tourism in Finnish Lapland, social inclusion has manifested itself over time. With long traditions, tourism in Lapland has grown to be one of the main livelihoods and has made indelible changes in small villages and municipalities, which are now more like associations of urban centres and rural villages.

In this chapter, inclusion is considered as providing possibilities for employment and improving the quality of life for people other than those working in tourism or related industries. All three businesses interviewed illuminated the subject from different angles.

Lapland as a Place of Residence and Tourism Destination, Rapid Tourism Development and High Unemployment Rate

Lapland is the largest province and also the most sparsely populated area in Finland: it covers 30% of the Finnish area, but with only 3% of the nation's population (Figure 6.1). This is illustrated best when comparing

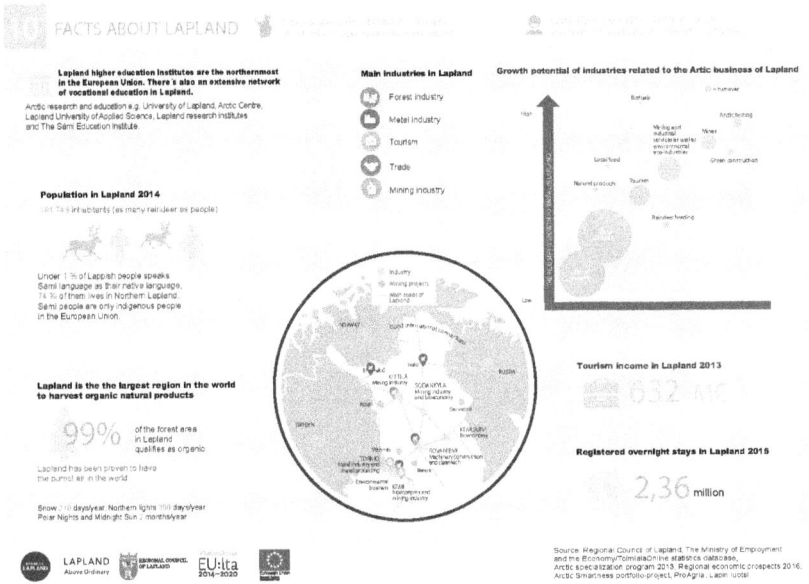

Figure 6.1 Ten facts about Lapland (House of Lapland, 2015)

population densities. In Finland, the average population density is 17.64 people/km², whereas it is only 2 people/km² in Lapland. In comparison, the population is most dense in the region of Uusimaa, where the capital of Helsinki is located, with 170 people/km²; the lowest population density is in Lapland, with municipalities such as Inari having only 0.45 people/km² (Statistics Finland, 2017).

Of the 21 municipalities of Lapland, only four are towns – Rovaniemi, Kemi, Tornio and Kemijärvi. All the other areas are rural and often ailed by the same problems as in many rural areas around Europe: an ageing population, a downturn in demographic trends and challenges related to employment and the provision of basic services.

The demographic trend in Lapland has seen a downturn since 1994. Up to 2016, the population in Lapland decreased by 10% while, over the past few years, the number of inhabitants has increased only in four Lappish municipalities. Although the employment rate in Lapland is approximately 2% lower than the Finnish average, there are large differences in unemployment rates between the different municipalities in Lapland, with eastern Lapland suffering the most from high unemployment. The number of people receiving income support is at the average national level in Finland, but costs resulting from social and health care services rank Lapland in the second highest position in Finland (Satokangas, 2018).

Apart from its sparse population, Lapland is also characterised by vast spaces and plentiful Arctic nature, thus making it an attractive tourist

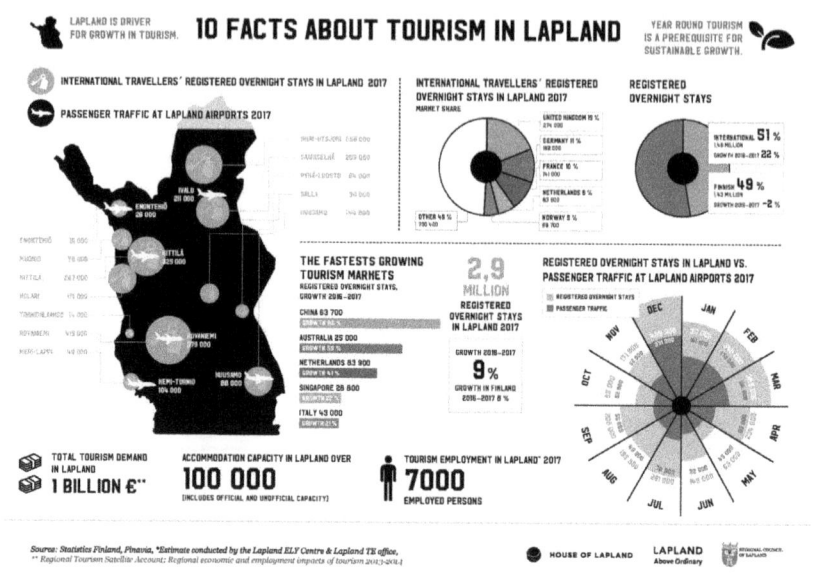

Figure 6.2 Ten facts about tourism in Lapland (House of Lapland, 2018)

destination. Tourism is one of the leading industries in Lapland and has grown rapidly over recent years, with the biggest increases in the winter season. The summer season has also seen an increase in overnight stays, but with less growth than the winter season. Lappish tourism attracts not only tourists but also investments. According to the Regional Council of Lapland (RCL), investments in the Lappish tourism sector were estimated to make up about 200 million euros in 2017–2018 (RCL, 2019a). Detailed data on tourism in Finland are provided in Figure 6.2.

The position of the tourism industry in Lapland's economic structure is now permanently rooted and the impacts of tourism are identifiable in many sectors. The sectors providing most employment in Lapland are health and social services, mining and related sectors, business services and retail (Lapland Above Ordinary, 2019). The private sector is currently the biggest employer in Lapland, with a total of 64%, in turn meaning that 36% of jobs are in the public sector (government, municipalities). Employment in the private sector is also growing, as structural changes have seen the creation of new vacancies while employment options have decreased in the public sector (Lapland Above Ordinary, 2019). The service sector has grown as a significant employer and still growing industries in Lapland are mining, tourism and services (Lapin Luotsi, 2019). The increasing use of temporary agency workers and human resources services are also offering new ways of finding job opportunities, while gained work experience leads to more stable and permanent contracts

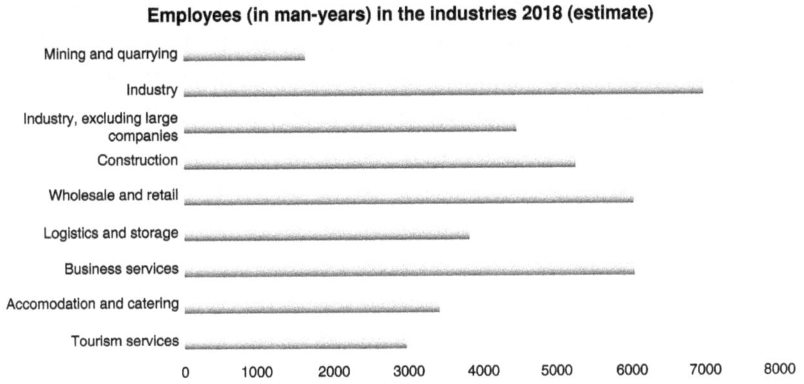

Employees (in man-years) in the industries 2018 (estimate)

Sector	Employees
Mining and quarrying	~400
Industry	~7500
Industry, excluding large companies	~4500
Construction	~6000
Wholesale and retail	~6000
Logistics and storage	~3000
Business services	~6000
Accomodation and catering	~2000
Tourism services	~1200

Figure 6.3 Number of employees in business sectors in Lapland (Lapin Luotsi, 2019 (translated and adapted by S. Nisula))

(TEM, 2019: 412). An overview of employment in Lapland is provided in Figure 6.3.

Some service industries have faced a lack of skilled workforce, as detected in the Occupational Barometer of Lapland (OBL). The OBL is an employment outlook covering all the economic regions of Finland. The OBL is updated twice per year and its main purpose is to increase the balance between job seekers and vacancies, as well as to promote occupational and regional mobility. The OBL is taken care of by Finnish public employment and business services, the Centre for Economic Development, Transport and the Environment (CEDTE) and the Ministry of Economic Affairs and Employment of Finland (TEM).

According to OBL (2019), there are several service industries directly related to the tourism industry in Lapland that are lacking workforce: hotel receptionists, travel guides, restaurant and catering workers, waiters and cleaners will be the most needed employees in the next six months in Lapland. In addition, it is noteworthy that numerous services linked to tourism industries are also facing a lack of job applicants (e.g. transportation and logistics). These estimates are published every six months, based on a view of the development of labour demand and supply in different sectors (OBL, 2019).

Another challenge is the mismatch between labour demand and supply. According to the Ministry of Economic Affairs and Employment of Finland (TEM, 2019), seasonal fluctuations in tourism and other industries (e.g. construction industry) impede the recruitment of skilled workers, causing challenges to the industry. During high season, there is not enough workers to satisfy the labour market needs and professional employees have to be recruited from outside Lapland. Likewise, the ageing and decreasing population in Lapland is a serious challenge that is affecting recruitment and workforce availability (TEM, 2019).

Efforts have been made to address the issue of employment mismatch and lack of workforce in tourism. For example, the national Matkailudiili project (translated from Finnish by the writers of this text as 'Tourism deal') has enabled quick pilot projects around Finland to help tourism companies recruit workers (CEDTE, 2019). In June 2019, the project Matkailun verkostomainen oppimiskenttä länsirajalla (translated as 'Networked learning field of tourism at the western border' by the writers of this text) began with the goal of remedying the lack of workforce and unemployment in the western Lapland area by mapping the needs of companies in tourism and related businesses and the existing know-how of unemployed people in the area. According to the project manager, the target group for actions in the project were people with no secondary education, of which men comprise about two thirds (Eura, 2019; Yle, 2019).

Social Inclusion in Tourism in Lapland: Three Different Perspectives

Let us recall the term inclusion and some of its characteristics. Some characteristics are the possibilities of having health, education, work, livelihood and social relationships. These can be viewed as the division of *having*, *belonging* and *acting*. Inclusion also means the possibilities and resources to enhance skills and capabilities. When considering these characteristics in Lapland, one might argue that some of these characteristics are missing for people other than, for example, the unemployed. For instance, access to health or educational services can be challenging in sparsely populated areas. Regarding health services in Lapland, topics of concern are the care of elderly people and the long distances expectant mothers have to travel to give birth (Maaseudun tulevaisuus, 2017). Rovaniemi and Kemi are the only places in Lapland with maternity wards: for some pregnant women, this means a four-hour drive. These types of health services fall into the dimension of having, as they concern well-being and safety. It is also possible that the dimensions of belonging and acting in an area suffering from outwards migration might not be completely realised.

In this study, our interviewees raised topics relating to all characteristics of inclusion – not only for those employed in tourism and related fields, but also for other people living in the area. The most visible parts of tourism are the 'exceptions' they cause to normal lives: passenger aeroplanes, tourist buses and tourists wandering around the area and taking part in (tourism) activities that local people usually do not. These are often services provided by tourism businesses for tourists. However, the importance of tourism also lies in the ripple effects it causes, which are not so obvious because they are part of people's everyday lives. These effects are not only economical, but they could also be considered cultural or social. The availability of a variety of such services and products that are also relevant for locals is one manifestation of this.

In this study, we asked about employment and the characteristic of inclusion was raised in the discussions. It was reported that the lack of workforce presents both a challenge and a possibility. Growth in tourism has been so strong that the industry has been battling with the problem of employment mismatch for some years and this is exacerbated by the still very seasonal nature of tourism in the region. In Lapland, winter sells. Efforts to put Lapland on the international tourism map began in full force at the turn of the millennium (RCL, 2017). Winter activities, Arctic nature, the northern lights and, of course, Santa Claus are the traditional explicit selling points of Lapland. All of these attractions belong more or less to the winter season (although you can still meet Santa Claus in the summer) and are in increasing demand. Branding of the summer of Lapland is a work in progress, as is the flux of summer tourists to Lapland.

Lack of workforce and different forms of employment

One might think that the seasonal nature of tourism would create unemployment rather than a lack of workforce. The problem of seasonality affects not only the tourism business, but is also a problem for individuals and the region. As already noted, growth in tourism in winter creates a high demand for workers in this season, with businesses having to hire people from outside Finland. When the winter business slows down, fewer employees are needed. Those workers coming from somewhere else then return home or go somewhere else to work. However, getting people to move to Lapland for just a season is not easy – Lapland is far away and experienced workers are more likely to be people with families and children: moving your whole family up north for a single tourist season might not be possible, especially if your spouse does not work in tourism.

This situation means that the need for workers presents opportunities for many types of employment. The VET college interviewed offers special needs education at the upper secondary level in over 20 towns and municipalities in Finland. Students of the college all have some type of special need in education or learning, which makes studying in a mainstream vocational college impossible for them. Efforts are made to provide every student with an individual path to a working life. The fields of education vary between towns; in Rovaniemi, students can acquire qualifications in culinary skills and catering or in cleaning and property services. The college does not offer a specific degree in tourism, but the college representative described both the college and students as benefitting from the growth in tourism. The VET provider's fields of education are needed in the tourism industry and the demand for workers opens up opportunities for students on many levels. Some students are at their best doing specialised and assisting tasks, and the workforce demand brings these opportunities to the job market. Due to the lines of education, suitable jobs in tourism for these students are mainly in restaurants, hotels, cleaning and property maintenance rather than in programme services.

Cooperation with tourism companies and related businesses already exists. College staff visit companies with the goals of establishing connections between college and work places and paving individual paths for students to working life. The college also has learning environment agreements with different enterprises, in which students and their supervisors study and practice tasks for a week in the company or in an event. The representative of the college interviewed called for incentives for businesses to hire partially disabled people. According to the interviewee, the success of a partially disabled person finding work depends solely on an entrepreneur's own particular will to hire partially disabled people. The students wish to do productive work and are diligent workers, but breaking into the job market can be challenging. This can be alleviated by taking part in regular activities in the society via working. This is not only important for everyone but it is also an economic issue for society: the college educates people using money from taxes – if the students cannot find paid work after graduating, they stay at home and need income from different types of subsidies (i.e. taxes).

In the current recently released government programme (Valtioneuvosto, 2019), increasing employment is one of the main goals and the biggest potential for achieving this increase is seen as the groups who are having difficulties finding work, such as partially disabled people. Our interviewee mentioned interesting models to increase the employment of partially disabled people existing elsewhere in Europe.

The other two interviewees (from Lappish tourism businesses) confirmed the same regarding job opportunities: the high demand for workers means that businesses are interested in all types of employment models. Both interviewees had experience of hiring people using different types of models, had good experiences and were open to the idea of hiring people in various ways in the future. One interviewee noted, however, that it is important for all counterparts to ensure the person coming to the company (for example to learn while working or on different types of apprentice or trainee programmes) has enough basic knowledge about the tasks involved so that the company would not have the full responsibility of the person's education on its shoulders. The interviewee from the VET college is trying to ensure this is the case by means of the work cycles of the staff and the learning environment agreements. With such means, everyone knows what type of work tasks are to be expected and what types of skills are needed.

Waiting for summer: The seasonal nature of tourism work

Room for growth in Lappish tourism still exists. As noted in the region's tourism strategy for 2020–2023 (RCL, 2020), tourism-related problems include seasonality, poor profitability and an underdeveloped business structure. These issues have created limited investments and new areas of business. Seasonality results in frequent changes of members of staff, a lack of permanent employment and low wage levels (RCL, 2020). The issue of sustainable, year-round tourism has thus been picked up as part of the region's tourism strategy vision for the year 2030: 'Lapland is a responsible and authentic year-round tourism destination, which grows sensibly' (RCL, 2020: 6). The strategy thus aims to support year-round tourism and increase capacity utilisation, which should improve profitability and reduce seasonal problems. It should also lead to the creation of new companies and their relocation to the area (RCL, 2020).

The interviewee located in the Fell Lapland area said the company employed 70 people year-round, with the number of employees doubling to 140 during the high season. The company's aim regarding social responsibility is to hire local people as much as possible. In this firm, the permanently employed people are all local, but about half of the seasonal workers have to be hired from elsewhere, which is symptomatic of the problem of the current mismatch in tourism employment. According to the interviewee, this presents problems on many levels, one of which is tax revenue going to the home municipalities of the visiting seasonal workers. In the case of the VET college, the ripple effects of tourism presented work opportunities for workers in associated fields. In terms of business, the ripple effects of seasonal workers mean a loss in tax revenue for municipalities.

One might think the problem is only economic. However, as we already noted, the concept of inclusion includes not only economy-related issues (such as work and livelihood) but also the dimensions of having, belonging and acting. With little tax revenue, municipalities can only provide limited services. In addition, seasonal work attracts people to the municipality for only short-term stays, and alone. The interviewed entrepreneur described utilising the summer period as a next major goal in tourism in Lapland as it would significantly benefit both businesses and the region. Tourism as an industry is a major player in bringing money and employment to the area. Year-round tourism would thus result in year-round employment for more people and thus increased tax revenue for the municipalities. One of the outcomes of this would possibly be a change in the direction of migration. According to one of the entrepreneurs interviewed, many seasonal workers coming from elsewhere could see themselves living in the area permanently, often with a tax-paying partner. In turn, more people living in the area would create more demand for different types of services, thus creating more work opportunities in the area. The need for public services would also increase, again bringing the possibility of more employees and more taxpayers. The tourism industry thus creates many different ripple effects, with extra side streams of income: it is estimated that one euro spent in tourism brings 56 cents to other businesses (Business Finland, 2018).

Keeping villages lively

According to the previously mentioned entrepreneur, tourism is a business that keeps sparsely populated and rural areas in Lapland alive. This is easy to believe since, according to this entrepreneur, already almost half of the municipality's turnover currently comes from tourism. This effect was also underlined by another interviewee, an entrepreneur from a small village in central Lapland, who described the village as being close to the brink of death before the start of the tourism business: agricultural activities had stopped, people were moving away and houses were abandoned. With the help of leader funding and in cooperation with the municipality, a review and strategy were conceived to revitalise the village. The entrepreneur brought cattle to the area and started to developed the tourism business around agriculture. The tourism business itself comprises a hotel and restaurant and employs three full-time employees; however, all of the farm work is subcontracted, so the total impact of the business on employment concerns 15–20 people. The amount of work does not vary between summer and winter, but the type of work does: during winter, the need for restaurant and hotel workers is higher, whereas there is more work on the farm in summer.

Farm work is bringing back demand for know-how on a subject already deemed not so necessary in the area and the ripple effects of the tourism business have brought more employment opportunities to the village. The interviewee noted that the launch of the business first brought domestic customers to the area. These people were cottage owners who started to spend more time at properties because there were now services available in the village. Since then, international tourists have also found the village and these constitute approximately 50% of incoming tourists. However, seasonal differences are seen also in this enterprise, with international tourists coming mostly during the winter.

The advent of this tourism business has also sparked a number of other new businesses in the area: at least two reindeer farms and a river ferry restaurant have since started, along with a few artisans. The entrepreneur also told how, since the kick-off of this business, the tradition of the village country fair had been revived, bringing at best 4000 guests to the village. Villagers can thus sell their products to visitors, utilise their skills and traditions, and learn what type of products sell. According to the entrepreneur, such sales are beneficial as they create income for the artisans and the village association, not just the interviewee's business. The interviewed entrepreneur has also started another business providing transportation for tourists.

Using other local services and businesses is a natural part of the actions of both companies interviewed. In addition to employing local people, both companies described using the products of local producers such as farmers and they also buy services from other businesses in the area (e.g. transportation and guides). Their businesses have created opportunities for other businesses and helped local people earn money from their expertise and products. With growth, these businesses may employ yet more people from the area.

Discussion

Tourism in Lapland often means tourism in rural, sparsely populated areas. Although Lapland has four towns/cities, the imagery that brings tourism to the area is that of nature, beautiful scenery and vast amounts of space as opposed to urban milieu and city life. However, outside the travel brochure imagery, these areas suffer from the problems typical of sparsely populated areas. Livelihoods previously practised no longer exist to the extent they used to and people are moving away to find opportunities and be part of a larger society.

An increase in tourism has evoked discussion about sustainability in arranging tourism activities. Ecological, economical and sociocultural sustainability walk hand in hand in this discussion, which also includes inclusion. From the viewpoint of sociocultural sustainability, both of the tourism business interviewees highlighted the importance of upholding local and Sámi cultures and traditions. This means at least two things – people can make a living out of many types of skills and traditions, and it happens on their own terms. Their traditional knowledge and skills are still valid in an urbanised world. Elements of Sámi culture have been used in the tourism industry in Lapland – unfortunately not always responsibly (Sámi Parliament, 2020). The Sámi Parliament (2018) has released an ethical guideline about tourism and Sámi culture. As of this writing, the guideline is yet to be translated into English.

The entrepreneur in Fell Lapland emphasised the importance of developing the summer season into an interesting and well-known product: this would mean more permanent work for people in the Lapland region and – as the entrepreneur underlined – it would make responsible growth in tourism possible. It would enable year-round tourism income and income from side streams. Reflecting on the effects of the desired year-round tourism to the larger phenomenon of inclusion, it would offer people working in a currently very seasonal tourism business the chance to settle instead of changing plans as the seasons change. This would also benefit the area, with more people living there and participating in its development both as taxpayers and as individuals with ideas and needs for the region. Seasonal workers come to the area, do their work and leave. Thus, one might argue that, for seasonal workers, the time they spend in the area is defined mostly by work. For people living in the area long term, work is not the only defining thing – free time and other aspects of everyday life come into play. More people need more services, which means more opportunities for the providers of services. Inclusion-wise, the three dimensions of inclusion – belonging, having, acting – could be realised.

When looking at the characteristics of inclusion, one might point out that, for seasonal workers who come from elsewhere, the criteria for inclusion – if we understand those characteristics as such – might not be met even though these workers have jobs and income. Housing is one problem for seasonal workers. They often live in cottages meant for tourism use or in some similar type of housing with other workers. The entrepreneur in Fell Lapland pointed out that even if a worker wanted to change their official place of residence to the municipality, it would not be possible because it is not possible to officially live in a holiday home. How included in the community seasonal workers currently feel is a question this text cannot answer; nor can we know if they have a sense of belonging, acting and having, and what these dimensions mean to them as seasonal workers. The new tourism strategy for Lapland (RCL, 2020) notes that many tourism destinations have been formed in areas where no local communities existed before. About his category of *being*, Allardt (1980: 46–47) states that the antithesis for self-realisation is alienation, which happens when your relations to your surroundings are defined mainly by the benefits you bring, for example as a worker or a customer. This is an interesting juxtaposition because, for a seasonal worker, the experience of working in Lapland might actually be comparable to the experience of being a tourist – and not just because both spend time on tourism during their stay. The participation of both in the local everyday life might be quite minimal and both are assessed by the economic benefits they bring to the area.

Since the growth in tourism, talks of aiming for even more growth can make people grin in scepticism or even in a little bit of fear. The initial vision might be that of even more people visiting already popular tourist destinations in the peak of high season. However, the entrepreneur in Fell Lapland pointed out that the potential of summer is still largely unused and harnessing the summer season could be done in a responsible way. Summer tourism would not require building more accommodation, but instead fully utilising the already existing capacity which is currently underused due to the quiet summer season. The crowded winter resorts would not become any more crowded. The entrepreneur also stressed that turning the summer season into a known product would require the organised actions of many stakeholders (including businesses, municipalities, tourism areas and the state) because making summer in Lapland known in the international market requires marketing a new product to new customers in an organised fashion. In addition, infrastructural and services-related challenges have to be solved to make year-round employment possible for more people. The new tourism strategy also underlines the role of the municipalities in organising construction and services in tourism areas (RCL, 2020). This would be enabled by making year-round life possible for not just young single people, but also for families.

In the interviews conducted for this work, the entrepreneurs described how full-time workers who have moved to the area have introduced new free-time activities to the area and how elderly people now the possibility to live longer in their own homes because local restaurants can provide lunch services now that there is demand for restaurant services in general. These examples illustrate how revitalisation and new opportunities are not limited to just the people employed in tourism – they are also for the community they live in. As pointed out by Grant and Kluge (2012), tourism can be a vehicle for the promotion of social inclusion through the facilitation of social interactions and networks, a widening of narrow social spheres, and consolidating identity as well as creating a new identity in the future. In sparsely populated areas with challenges in the provision of basic services, these types of cross-overs between seemingly different fields of tourism and care could be worthy of further exploration.

Another aspect of the tourism business is the demand for authentic experiences. In the case of the Arctic regions, the question of safety is also noteworthy. Seminars on the future of tourism discuss to what extent issues like robotics and artificial intelligence will be applied in tourism. However, currently, many of the tasks in tourism and related businesses are those where humans might be difficult to replace, especially with respect to authenticity and safety in the Arctic environment. Of course, the future remains to be seen, but tourism might be an industry that is not the first to be overtaken by robots.

The vocational college actively communicates its actions through its website as well as to its stakeholders in everyday operations. The tourism company in Fell Lapland communicates its strategies in responsibility and sustainability on its website and makes sure its employees know about the ways the company works, but say they could probably communicate more about the aspect of inclusion. The company in Central Lapland also said its does not communicate about these things particularly – it just carries out its business. From the interviews, sustainability came through as more of an internalised mind-set than something that is actively highlighted or, in a way, performed.

Although the growth in tourism has led to a lack of workforce and thus possibilities for many types of employment models, on the other side of the coin lies the question of what happens when growth stabilises. As of this writing, the new Finnish government had just released a government programme titled 'Inclusive and competent Finland: A socially, economically and ecologically sustainable society' (Valtioneuvosto, 2019), which sets the explicit goal of transforming Finland into a socially sustainable society. One of the goals is growth in employment and the biggest potential is seen as groups currently having difficulties in finding work, including older people and people with partial work capacity. Tourism is recognised as an important industry. Developing sustainable tourism and adventure services are seen as opportunity in overall development. In addition, a national programme is planned for the tourism sector to support sustainable growth of the sector. The objective of basing the development on regional strengths is set with various measures, along with recognition of the special considerations needed in sparsely populated areas.

Considering the economic significance of the tourism sector in the Lapland region and the challenges Lapland is facing as a sparsely populated area, the government programme raises optimism in considering employment, quality of life and inclusion. Of course, the government programme addresses other issues as well on its over 220 pages and the realisation of these goals will depend not only on what is written in the programme. However, as the representative of the vocational college noted, examples for models that strive to advance the employment of partially disabled people can be found from other countries. Governmental or legislative routes are often long and not very agile, but they are also not the only ones available. A set of strategies exist in Lapland to support the vitality of the area. The regional development plan describes the desired direction for the development of the whole region and the tourism strategy pinpoints the areas of development for tourism, with growth and competitiveness as a goal (RCL, 2019a, 2019b). Lapland is already a participating region in the thematic area for social economy in the smart specialisation platform of the European Commission (European Commission, 2019). The new tourism strategy emphasises responsibility in tourism services (RCL, 2020) and the current regional development strategy spans the years 2018–2021. It remains to be seen how the aforementioned points from the government programme and smart specialisation manifest themselves in the new strategy.

In view of these findings and discussions, more research on social inclusion in tourism should be conducted. With Lapland participating in the European Commission's S3 Social Economy Thematic Platform, it could be investigated whether development on a strategic level has a concrete impact on tourism enterprises and its employment practices. Likewise, future research could explore whether the growing tourism sector in Lapland and with it its growing labour demand will lead to a wider spread of social inclusion practices in tourism enterprises employment processes.

Conclusion

Social inclusion has been deemed to be an important part of the fabric of welfare states such as Finland. With social inclusion being a vital part of sustainable development, its growing importance is also recognised in

Lapland. As the tourism industry keeps growing, the demand for labour is also on the rise in Lapland. Hence, a possibility exists for those in a difficult labour market position to be included in the momentum provided by the accelerating tourism industry growth, therefore narrowing the gap between workforce supply and demand.

Nevertheless, although social inclusion was observed in the frame of the interviews conducted for this research, it is not yet an integral part of the enterprises' communications, unlike in the examples from other European regions examined in the introduction of this chapter. In the enterprises interviewed here, social inclusion seems to happen naturally through the employment of locals, which anchors them to the area and avoids the emptying of already sparsely populated areas. In addition, more institutionalised approaches through social inclusion in the form of wage subsidies and work placements have been implemented by the enterprises. With the increasing importance of sustainability in tourism from customers' perspectives, the communication of social inclusion principles and activities might be worth a try.

References

AFW (Association for Finnish Work) (2019) Key flag. See https://suomalainentyo.fi/en/services/key-flag/ (accessed March 2020).

Allardt, E. (1980) *Hyvinvoinnin ulottuvuuksia*. Porvoo: WSOY.

Brownies and Downies (2019a) Geschiedenis. See https://www.browniesanddownies.nl/geschiedenis (accessed March 2020).

Brownies and Downies (2019b) Partners. See https://www.browniesanddownies.nl/partners (accessed March 2020).

Business Finland (2018) Tourism as export infographic 2018. See https://www.businessfinland.fi/globalassets/new-pictures/talent-boost/tourism-2018-as-export-infographic.pdf (accessed March 2020).

Business Finland (2019) Vastuullisuus. Kestävyys matkailuvalttina. See https://www.businessfinland.fi/suomalaisille- asiakkaille/palvelut/verkostot/matkailu/vastuullisuus/kestava-matkailu-lyhyesti/ (accessed March 2020).

CEDTE (Centre for Economic Development, Transport and the Environment) (2019) Matkailudiili. See http://www.ely-keskus.fi/web/ely/pohjois-pohjanmaa- matkailudiili%3Bjsessionid=43573A9183CE566624DC50ACA9888CFE?p_p_id=122_INSTANCE_aluevalinta&p_p_lifecycle=0&p_p_state=normal&p_p_mode=view&p_r_p_564233524_resetCur=true&p_r_p_564233524_categoryId=14404 (accessed March 2020).

Dreher, M.T., Carrion, R.D.S.M. and da Silveira, A.P.K. (2013) Working in the tourism sector: social inclusion and prejudices. *Cuadernos de Turismo* 32, 281–294.

Eura (2019) Matkailun verkostomainen oppimiskenttä länsirajalla. See https://www.eura2014.fi/rrtiepa/projekti.php?projektikoodi=S21587 (accessed March 2020).

European Commission (2019) Social economy in the EU. See https://ec.europa.eu/growth/sectors/social- economy_en (accessed March 2020).

Grant, B.C. and Kluge, M.A. (2012) Leisure and physical well-being. In H. Gibson and J.F. Singleton (eds) *Leisure and Aging: Theory and Practice* (pp. 129–142). Aging and Mental Health. Champaign, IL: Human Kinetics.

Harju-Myllyaho, A., Kohllechner-Autto, M. and Nisula, S. (2017) Study on the situation and the legal framework of social entrepreneurship in Lapland, Finland. Rovaniemi: Lapland University of Applied Sciences.

House of Lapland (2015) Infographic: 10 Facts about Lapland. See https://www.lapland.fi/business/facts-figures/infographic-10-facts-about-lapland/ (accessed March 2020).

House of Lapland (2018) Infographic: 10 Facts about Tourism in Lapland. See https://www.lapland.fi/business/facts-figures/infographic-10-facts-tourism-lapland-2017/ (accessed March 2020).

ICA (International Cooperative Alliance) (2016) Cooperative identity, values and principles. See https://www.ica.coop/en/cooperatives/cooperative-identity (accessed March 2020).

Kohl, J. (2006) Working together for growth and jobs. A new start for the Lisbon strategy. Sosiaalinen ulottuvuus tulevaisuuden voimavarana. Avauksia rajapintakeskusteluille. Sosiaali- ja terveysministeriön julkaisuja. Helsinki: Yliopistopaino

Lapin Luotsi (2019) Elinkeinorakenne. Lapin toimintaympäristö, elinkeinot ja aluetalous, elinkeinorakenne ja talouskehitys. See http://luotsi.lappi.fi/elinkeinorakenne (accessed March 2020).

Lapland Above Ordinary (2019) Lapin suhdannekatsaus 2019. See http://luotsi.lappi.fi/c/document_library/get_file?folderId=683161&name=DLFE-34910.pdf (accessed March 2020).

Leemann, L., Kuusio, H. and Hämäläinen, R.-M. (2015) Sosiaalinen osallisuus. Sosiaalisen osallisuuden edistämisen koordinaatiohanke (Sokra). Terveyden ja hyvinvoinnin laitos. See https://thl.fi/documents/966696/3775621/Tietopaketti_Sosiaalinen_Osallisuus.pdf/4bc56a65-8eb2-41c3-87b8-0cd963a2c600 (accessed March 2020).

Maaseudun tulevaisuus (2017) Pelottaako, että tulee matkasynnytys? – Lue Lapin äitien kokemuksia. See https://www.maaseuduntulevaisuus.fi/ihmiset- kulttuuri/pelottaako-ett%C3%A4-tulee-matkasynnytys-lue-lapin-%C3%A4itien-kokemuksia-1.197864 (accessed March 2020).

Magdas Hotel (2019) See https://www.magdas-hotel.at/en/ (accessed June 2019).

NIHW (National Institute for Health and Welfare) (2018) Sosiaalinen osallisuus teoreettisena käsitteenä. See https://thl.fi/fi/web/hyvinvoinnin-ja-terveyden-edistamisen-johtaminen/osallisuuden-edistaminen/heikoimmassa-asemassa-olevien-osallisuus/mita-sosiaalinen-osallisuus-on-/sosiaalinen-osallisuus-teoreettisena-kasitteena (accessed March 2020).

NIHW (2019) Osallisuuden osatekijät. See https://thl.fi/fi/web/hyvinvointi-ja-terveyserot/eriarvoisuus/hyvinvointi/osallisuus/osallisuuden-osatekijat

OBL (Occupational Barometer of Lapland) (2019) Työllistymisen näkymät eri ammateissa. See https://www.ammattibarometri.fi/vertailu.asp?maakunta=lappi&vuosi=19i&kieli= (accessed March 2020).

Polese, M. and Stren, R. (2000) *The Social Sustainability of Cities: Diversity and Management of Change*. Toronto: University of Toronto Press.

Raivio H. and Karjalainen J. (2013) Osallisuus ei ole keino tai väline, palvelut ovat! Osallisuuden rakentuminen 2010-luvun tavoite- ja toimintaohjelmissa. In Taina Era (ed.) *Osallisuus – oikeutta vai pakkoa?* (p. 12–34). Jyväskylä: Jyväskylän ammattikorkeakoulu.

RCL (Regional Council of Lapland) (2017) Matkailun menestystarina. See https://issuu.com/lapinliitto/docs/matkailunmenestystarina (accessed March 2020).

RCL (2019a) Matkailu. See http://luotsi.lappi.fi/kuvaajat-matkailu (accessed January 2021).

RCL (2019b) Lappisopimus. See http://www.lappi.fi/lapinliitto/lappi-sopimus (accessed March 2020).

RCL (2020) Lapin matkailustrategia 2020–2023. See http://www.lappi.fi/c/document_library/get_file?folderId=17957&name=DLFE-35907.pdf (accessed March 2020).

Sámi Parliament (2018) Vastuullisen ja eettisesti kestävän saamelasimatkailun toiminta-periaatteet. See https://www.samediggi.fi/wp-content/uploads/2018/11/Vastuullisen-ja-eettisesti-kest%C3%A4v%C3%A4n-saamelaismatkailun-toimintaperiaatteet_hyv%C3%A4ksytty_24092018-3.pdf (accessed January 2021).

Sámi Parliament (2020) Culturally responsible Sámi tourism. See https://www.samediggi.fi/ongoing-projects/culturally-responsible-sami-tourism/?lang=en (accessed March 2020).

Satokangas, P. (2018) Arjen turvan toimintaympäristö Lapissa - alueellisten haasteiden ja voimavarojen kuvailua. In N. Niemisalo (ed.) *Arjen turvaa Lapissa. Osallisuus, palvelut ja elinkeinot* (pp. 29–37). Rovaniemi: Lapland University of Applied Sciences.

SOCENT SPAs (2017) Social Entrepreneurship in Sparsely Populated Areas. Project summary. https://www.interregeurope.eu/socentspas/

TEM (Ministry of Economic Affairs and Employment of Finland) (2019) Alueelliset kehitysnäkymät kevät 2019. See http://julkaisut.valtioneuvosto.fi/bitstream/handle/10024/161538/TEM_28_19_Alueelliset_kehitysn akymat_Kevat_2019.pdf (accessed March 2020).

Universo Santi (2019) See http://universosanti.com/proyecto/#accion (accessed June 2019).

UNWTO (United Nations World Tourism Organization) (2020) Sustainable development. See https://www.unwto.org/sustainable-development

Valtioneuvosto (2019) Inclusive and competent Finland: A socially, economically and ecologically sustainable society. See http://julkaisut.valtioneuvosto.fi/bitstream/handle/10024/161664/Inclusive%20and%20competent%20Finland_2019.pdf?sequence=7&isAllowed=y (accessed March 2020).

WTTC (World Travel & Tourism Council) (2020) UN Sustainable Development Goals. See https://www.wttc.org/agenda/un-sustainable-development-goals/ (accessed March 2020).

Yle (2019) Länsirajalla ratkotaan matkailun työvoimapulaa – työpaikka voi löytyä hyvinkin vähällä koulun penkillä istumisella. See https://yle.fi/uutiset/3-10839087 (accessed March 2020).

7 Conclusion

Anu Harju-Myllyaho and Salla Jutila

When we started this book process, our aim was to gain a better understanding of the future of tourism inclusion, to prompt questions and to increase curiosity on this issue. Furthermore, we wished to present some critical factors concerning the future of inclusive tourism, as well as practices and methods that might work as building blocks to pave the way towards more inclusive tourism futures. The chapters in this book support the understanding of inclusion in tourism as a multifaceted concept. The foundations of inclusion are openness, transparency, equality and freedom of choice – including the choice to *not* be included. These are the core principles that guide our actions in the development of inclusive tourism. In this concluding chapter, our aim is to reflect on the important viewpoints presented in the chapters from the perspective of future research and foresight.

In practice, one of the significant factors in developing inclusive tourism is recognising different tourism actors or stakeholders. This is tricky because tourism is an industry that pierces society in a unique way and touches people's lives even if they do not operate in the industry. It is even trickier if we also consider the different drivers of change and how they impact our future. However, this is also something about which we can have a say ourselves, considering that the future does not come as a given – it is made.

Understanding the Changing Environment: Megatrends and Tourism Inclusion

The chapters of this book give a good idea of the issues that need to be taken into consideration if we want to develop inclusive tourism. At the same time, we must be able to understand environmental change. At the beginning of this new decade, various reports on trends were published that shed light on what could be in store. This changed rapidly with the COVID-19 crisis. Before the pandemic shook the whole world, the Finnish innovation fund Sitra published a report on the megatrends that will shape the years to come. The report states that exploring megatrends is an important part of futures thinking. A megatrend is a multidimensional range of drivers of change that offers a broader perspective on the future. The future is often thought to be a continuation of the present and megatrends thus offer a glimpse of where we might be heading (Dufva, 2020: 3). Sitra has published megatrend reports before, and Dufva (2020: 6) notes that the megatrends have not changed – at least not significantly – since the previous report. Certain elements are given more emphasis and futures thinking has developed; it is important to know about futures, to know how information is used and to be aware of different outlooks, so that existing assumptions can be challenged (Dufva, 2020: 7). This is exactly what the chapters in this book provide – a glimpse into possible future trajectories.

According to Dufva (2020: 9) the most profound change-maker is our response to ecological crises. In this respect, the COVID-19 pandemic brought an unexpected turn. However, as noted, megatrends are wide phenomena that include various parallel trends and even anti-trends; this is why we should treat megatrends as continuous change-makers, even though a change in the form of a global pandemic is now disturbing the way we see and are in the world. Dufva observes that we need to start ecological rebuilding soon – the next 10 years will be crucial in terms of how we succeed in building a resilient society that is able to adjust to climate change. Another significant development, according to Dufva, is that the division of power is shifting from a multinational world to one where the degree of economic, technological and cultural interaction and the width of networks define power. Societal systems are being put to the test as they try to offer a way to solve ecological crises and meet the challenges of a complex world. At the global level, strong leaders are expected to bring clarity to the confusion but, at the same time, there is a growing demand for change at the grassroots level (Dufva, 2020: 9).

This trend is also being seen in tourism, where the social aspect of sustainability is gaining in importance and new actors are emerging and making their voices heard. They are certainly needed. Dufva (2020: 9) asks how we can change culture, social structures and the educational system. Inclusive development is one way to tackle these challenges. By shifting power to various actors at different levels, society will – hopefully – become more robust.

Economic development, which causes tension in some societies, will play a significant role in the future of inclusion. In a way, people do have a common understanding of environmental challenges and the polarisation of economic well-being, but they also face difficulties in changing their habits (Dufva, 2020: 9). In tourism, this raises the question of who is entitled to travel if, on one side of the coin, there is the environmental crisis and, on the other, global international tourism exports are worth US$1.7 trillion (UNWTO, 2020: 2). The question of who is entitled to travel has also gained new dimensions as a result of the COVID-19 crisis. Images of people travelling have evoked both hope and critique in terms of tourism. This is due to the common belief that we are entitled to travel opposed to the fact that we should restrict ourselves from these pleasures of the modern Western lifestyle for the common good.

In tourism, the complexity of the situation is reflected in trends (e.g. Yeoman & McMahon-Beattie, 2019). In addition to megatrends, a plethora of trends (microtrends) and weak signals can help us to navigate the future. Microtrends are consumer-driven. According to Yeoman and McMahon-Beattie (2019: 115), there are individuals who have not had a once-in-a-lifetime experience. But, people are also now living longer and many of them have the chance to experience the same or similar things more than once. Thus, they might marry more than once or have *several* once-in-a-lifetime vacations. As Yeoman and McMahon-Beattie (2019: 15) observe, people have the time to complete more than one bucket list. Another microtrend mentioned by Yeoman and McMahon-Beattie is the luxury experience, which has become more mainstream. At the same time, the notion of luxury is being transformed. It is now more flexible and experiences have taken centre stage even if, for some, luxury still means expensive, high-quality products and services. The acquisition of skills, personal growth and cultural awareness has become a part of the luxury experience. Consequently, it is reasonable to expect that inclusive tourism will become something that tourists will demand and appreciate. They are also seeking authenticity in products and services that are unique to a region or a country, and they want those products and services to be ethical (Yeoman & McMahon-Beattie, 2019: 20).

Technological development is also mentioned as a microtrend by Yeoman and McMahon (2019: 20). This shifts the power from travel agencies to tourists, but this process is not straightforward. As Fernández-Villarán *et al.* pointed out in Chapter 5, the information needed to be able to make a travel decision is not the same for everyone. A tourist with special needs has to obtain sufficient information to decide whether a destination or service is suitable for them. This is not always possible using websites alone, which can be problematic from the tourist's viewpoint.

Megatrends and smaller, even local, trends provide interesting insight into the future of inclusive tourism. There seem to be contradictory streams of development. On the one hand, there is a clear demand for more sustainable products and services. On the other, as has been noted above, some consumers – including experienced tourists – are unable (or are unwilling) to change their behaviour. At the same time, the pace of tourism is increasing as people live to be older and new groups are (financially) equipped to travel. Are they prepared to make sacrifices for the sake of the environment and ethics?

Another interesting contradiction has been created by technological development. On the one hand, it advances inclusion by providing platforms for involvement. On the other, it creates tension and new forms of segregation between those able to take part in producing and consuming tourism and those who are, in one way or another, unable to do so (see, for example, Chapter 2 of this book). According to the tourism intelligence company Skift (2020: 17), in Western Europe, the public sector has begun to use digital platforms to gather feedback from local stakeholders in tourism. Thus, digital platforms create spaces where stakeholders can be engaged in decision-making. However, in terms of inclusion, as we have discovered in this book, it is very important that stakeholders feel they truly have an impact and that their voices are heard. As has been pointed out, the participatory approach can be a successful means of including local residents and other stakeholders in the development of tourism. Online platforms may face challenges in terms of inclusion, but digitalisation is transforming services. Providers now have to meet the needs of individual travellers.

Economic development is a key aspect of inclusive tourism. Consulting company PricewaterhouseCoopers (2020) describe it as being one of the key drivers of change. Economic advances in emerging markets are clearly important in whether habits of travel are encouraged or restricted for certain groups. Economics also decides who has the resources to influence decision-making and where the economic effects of tourism are channelled. Growing inequality will bring about many challenges to inclusive tourism because it has an impact on everyday life. Kraus *et al.* (2017: 422) suggest that societies and individuals suffer as inequality increases. As growing inequality has an influence on people's lifestyles, it will surely affect tourism as well.

Looking back to history might also provide a chance to understand the drivers of change and events that change the course of development. According to Yeoman and McMahon-Beattie (2020: 2), past events and turning points can be used as tools for anticipating the future. One example of this can be seen in the COVID-19 crisis. Scholars have turned their focus to past events, times of crisis and other pandemics. These can provide ideas for alternative paths from the crisis and help to anticipate the impact of different decisions and post-crisis development.

The Silenced, the Acknowledged and the Empowered

Canosa and Moyle (2016) summarise research related to the social impacts of tourism and note that it is a significant part of tourism research, but also a target for critique because it takes a narrow perspective on tourism stakeholders. They provide an important aspect to the discussion by bringing up the viewpoint of youth, whose voices have not been heard except in market and consumer research. This viewpoint is important because it pinpoints that there might be other important groups, stakeholders and actors that might be left on the margins. Different methods should also be taken into consideration to ensure these voices are heard (Canosa & Moyle, 2016).

The voice of the youth is a timely theme, particularly since Swedish youngster Greta Thunberg started a school strike for climate change. Thunberg, along with her followers, has raised diverse discussion with both critical and supportive discourses. The conversation has been filled with comments about Thunberg's youth and health. Her position is indeed interesting, because she is also a young girl and has Asperger's. When building paths towards more inclusive futures, the structures that are formed by the drivers of change should be taken into consideration, alongside what kind of subject positions are enabled by these structures.

Canosa and Moyle (2016) surveyed tourism research concerning children and youth and, applying the classification of Nielsen and Wilson (2012), classified 30 studies into categories of *silent*, *acknowledged* and *youth-oriented*. Nielsen and Wilson, meanwhile, presented an indigenous tourism typology with four positions: *invisible*, *identified*, *stakeholder* and *Indigenous-driven* (Nielsen & Wilson, 2012). Similar classifications can be applied in many instances and with different groups and, by doing so, it is possible to draw a picture of how inclusion is realised in research, planning and development. However, it is very important to take into consideration the timely trends, drivers of change and development paths – we must determine what these changes mean from the perspective of tourism inclusion and how we can impact them.

What enables (or hinders) development are, for instance, social structures that might be difficult to recognise in our daily lives. By making visible the structures that have an impact on participation – such as social class, culture, ethnic background and gender – it may be possible to identify paths to alternative future states in which these structures do not restrict action. Practices and activities are the visible form of inclusion. Many practices have been described in this book, including planning and strategy work, projects, pro-poor tourism activities and social entrepreneurship. Bringing to the foreground good practices and reorganising the not so good not only helps those involved but also provides models for future activities and development.

Intersectionality has been seen as a theory for producing information on different subject positions, which makes it possible to recognise different groups that have – for some reason or another – been neglected or oppressed and their special positions in social structures, to scrutinise these positions and the experiences they produce and make them visible (see Crenshaw, 2006). Intersectionality has also been named as one of the central ideas of the Finnish government's equality programme, which is a sign that this line of thinking is gaining ground on a wider scale.

Towards Inclusive Tourism Futures

Inclusive tourism as a concept might sound idealistic. However, there are concrete possibilities behind this beautiful idea. In terms of the future of inclusive tourism, the strength of the industry lies in the hospitality that is formed between hosts and guests (Nousiainen, 2015: 10). This involves welcoming people as respected customers, colleagues and co-operators. Hospitality, put very simply, is a way of being with the other (e.g. Chapter 1 in this book; Höckert, 2015: 91–136). In Chapter 3 of this book, Harju-Myllyaho and Jutila claim that the notion of hospitality, in a way, does not make assumptions concerning the participants and their needs. In this respect, inclusive tourism cannot exist without hospitality: hospitality is a central element of inclusive tourism.

Tourism is at the forefront of globalisation and international movement. It increases cultural awareness and thus supports the inclusion of people of different backgrounds. More people can now afford to travel – an activity that was previously considered a luxury. There has been a shift from the west to the east, and we expect that non-Western worldviews will gain ground in the industry. This means that tourism will become richer in cultural content and the shift will also secure economic growth. Having said that, we are aware that the chapters in this book are centred on Europe and thus do not geographically represent the globe. However, we encourage the reader to consider various geographical and cultural dimensions of inclusive tourism.

Supporting the idea of inclusive tourism on a more profound level, there is also the notion of freedom of movement, which is recognised in the Universal Declaration of Human Rights (United Nations, 1948). That said, the COVID-19 pandemic has created restrictions in movement that were unthinkable before spring 2020.

Tourism is a complex industry comprising vast networks. It is a vital component of various other industries and it is naturally affected by environmental change. At the time of writing, a number of sudden events had recently taken place that created shockwaves in the industry: the somewhat unexpected bankruptcy of Thomas Cook and the COVID-19 pandemic are two examples. These events have presented challenges, but the tourism business has often proven itself to be resilient. Networking, creativity and solution seeking – all features of the industry – are important in a world that is being continuously transformed.

In terms of inclusive tourism, a lack of development resources must be one of the key issues. A hectic environment affords little time for future-oriented thinking and development activities. Fortunately, this is gradually changing, and companies are paying more and more attention to sustainability, including social sustainability, and consequently to inclusive tourism.

However, despite the sublime goal of inclusive tourism, tourism remains exclusive in nature because there are and will be groups who, for one reason or another, are not able to travel. There will also be people or groups who are not recognised by the industry and who cannot get their voices heard. Restrictions on taking part in tourism might be social, political, economic or physical, or they might have to do with communication. For instance, for people with disabilities, a lack of information leads to uncertainty about service quality and suitability (see Chapter 5 in this book). In practice, inclusivity depends on the accessibility, openness and clarity of information.

Outcomes of the COVID-19 pandemic might be that some groups are overlooked by the industry or that the general mindset towards global industries and free movement are seen more critically and from different perspectives. For instance, in this pandemic, local people in Finnish Lapland felt threatened by travellers from the south because – even though the health care system in Finland is excellent – the health care system in Lapland might not be able to handle masses of travellers should many of them need intensive care because it was designed for the less than 200,000 inhabitants of Lapland. Social media was filled with messages from locals asking travellers to come back later, when it would be safer. The message 'dream now, travel later' was seen around the world. If we consider inclusive tourism, this situation brings forth a contradiction – on the one hand, we have the right to travel; on the other, locals also have the right to their well-being, health and neighbourhoods.

There are other possibilities in inclusive tourism: new business models to explore (although these new models might conflict due to unequal shares of services and income) and new target groups, actors and stakeholders that we need to be aware of. People's values are changing and customers are demanding more sustainable services as consciousness grows. From a broad perspective, inclusive products and services can mean access to more groups and more involvement by local residents, who can also have their voices heard and share the benefits of tourism.

Although the industry is paying more attention to sustainability and inclusion, certain trajectories need to be followed because they might have a negative effect on its development. As Dufva (2020) pointed out, the ecological crisis is one of the most pressing issues of our time. However, sustainable development needs to be considered not solely from an ecological perspective; its different aspects are connected and thus ecological sustainability cannot advance if the social dimension is not given its due. Similarly, inclusivity will be threatened if we are not able to find solutions to the ecological crisis. The ecological crisis is also a human rights crisis, as has been pointed out by various sources such as the United Nations (2015). In this respect, climate change might bring along challenges that we have not yet thought of. How, then, are we to consider the unthinkable? Asking the 'what if?' questions is one way of preparing for uncertainty and the kind of events that might catch us by surprise. Even though we cannot predict the future, we can consider alternative futures and the kinds of possible outcomes that we do not expect to become reality.

As has already been established, tourism is affected by global trends, and growing economic inequality is one of its principal challenges.

Increasing polarisation and segregation are other significant trajectories that threaten inclusive tourism. They will impact the way difference is accepted, who is considered and welcomed as a guest and who has enough capital to be the host. Yet, alongside awareness, methods and means are advancing. Participatory approaches to development and policymaking support companies and tourist areas in taking steps towards greater inclusivity. Inclusive policymaking in this sense means more than just including different stakeholders in policy papers; it also means involving them directly in the policymaking process from the beginning. This will empower new actors and aid in the inclusive development of providers of future tourism, as well as helping to accommodate the widest range of customers. Social entrepreneurship is a practice that can provide answers to many of the challenges faced by the industry. Entrepreneurship is changing along with tourists' values and demands. Social enterprises aim not only to do business, but also to support the prosperity of others. Given current developments and trends, we expect to see more innovative businesses and services with social elements. Examples of this have been presented elsewhere in this book (Chapter 6).

Modelling Inclusive Tourism Development: Building Inclusive Tourism Futures

In light of the contributions in this book, we suggest that the future of inclusive tourism could be built on the following components: actors, methods and practices. By adapting these elements, we have compiled a model (see Figure 7.1) that, inspired by the authors and chapters of this book and drawing from futures perspectives, current trends and development streams, could work as a general guideline as we build more inclusive tourism futures. The essential elements for the future of inclusive tourism are listed below, but this list is not exhaustive.

- Recognising and deconstructing personal and social norms and thought patterns.
- Recognising and anticipating various stakeholder groups (again, not an exhaustive list, but could include the silenced, the acknowledged, the empowered)
- Exploration and critical examination of methods and viewpoints; producing (futures) knowledge (futures research and foresight methods); participatory approaches; other novel and critical methods; discerning and critically scrutinising different futures and drivers of change, taking into consideration both micro and macro environments.
- Dialogue between different actors.
- Acknowledging the significance of everyday choices.

The model shown in Figure 7.1 is based on the active deconstruction of our patterns of thought and uses various methods that would help to

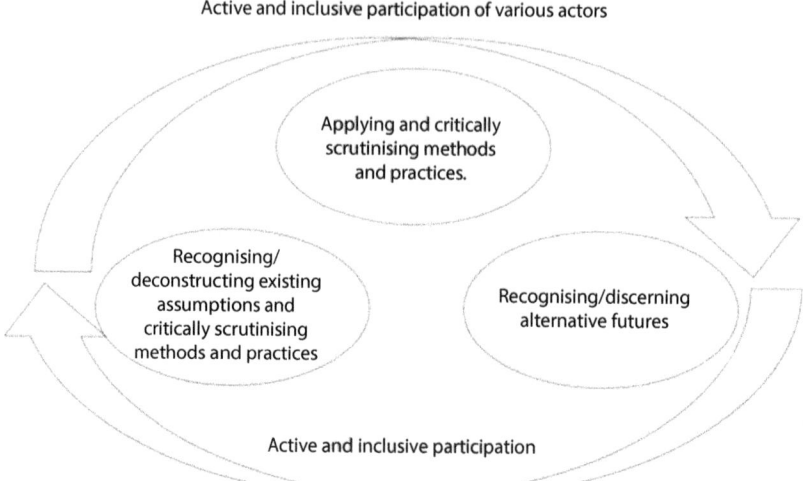

Figure 7.1 Building inclusive tourism futures

build alternative futures or futures images that will continue producing information for us to use in the present. The process is continuous – a loop that produces new information, inclusive practices and alternative futures images.

Inclusive Tourism in a World of Uncertainties and Contradictions

The future of inclusive tourism is uncertain and the industry faces a number of contradictions. Firstly, a dichotomy exists between the exclusive nature of tourism and individual experiences and the desire to enable tourism for all. Secondly, there is a need to reduce emissions and the undesirable effects of tourism while providing services that enable more people to travel and enjoy touristic experiences. Finally, there is a tension between the right to participate in tourism and the realistic likelihood of being part of it as a service producer or a customer. In an industry full of possibilities, restrictions and contradictions, there is no single answer to the question of futures-inclusive tourism and how it should be developed. However, if we want to make the industry inclusive, we must acknowledge different ways of being, doing and knowing.

There is not one future but many possible futures. It could be claimed that the future is decided. In practice, this involves the small choices that, in the field of tourism, are made regarding product development, marketing, digital platforms and in face-to-face contact. It is also about greater lines – adding more actors, expertise and knowledge can make a positive impact on those lines. Through the complex web of these processes it is possible for tourism to reorganise itself to enable more inclusive futures.

References

Canosa, A. and Moyle, B. (2016) Can anybody hear me? A critical analysis of young residents' voices in tourism studies. *Tourism Analysis* 21, 325–337.

Crenshaw, K. (2006) Mapping the margins. Intersectionality, Identity politics and violence against women of color. *Kvinder, kön & forskning* 2/3, 7–20.

Dufva, M. (2020) Megatrendit 2020. See https://media.sitra.fi/2019/12/15143428/megatrendit-2020.pdf (accessed January 2021).

Höckert, E. (2015) Ethics of hospitality: Participatory tourism encounters in the northern highlands of Nicaragua. PhD dissertation, University of Lapland.

Kraus, M.V., Park, J.W. and Tan, J.J.X. (2017) Signs of social class: The experience of economic inequality in everyday life. *Perspectives on Psychological Science* 12 (3), 422–435.

Nielsen, N. and Wilson, E. (2012) From invisible to indigenous-driven: A critical typology of research in indigenous tourism. *Journal of Hospitality and Tourism Management* 19 (1), 67–75, doi:10.1017/jht.2012.6.

Nousiainen, J. (2015) *Vieraanvaraisuuden käsitteellisiä ja paikallisia ulottuvuuksia. Mistä on lappilainen vieraanvaraisuus tehty -selvityshankkeen raportti*. Rovaniemi: Matkailualan tutkimus- ja koulutusinstituutti.

PricewaterhouseCoopers (2020) Shift in global economic power. See https://www.pwc.co.uk/issues/megatrends/shift-in-global-economic-power.html (accessed January 2020).

Skift (2020) Megatrends defining travel in 2020. See https://skift.com/2020/01/07/the-megatrends-defining-travel-in-2020/ (accessed January 2020).

United Nations (1948) Universal Declaration of Human Rights. See https://www.un.org/en/universal-declaration-human-rights/ (accessed March 2020).

United Nations (2015) Climate change and human rights. See https://wedocs.unep.org/bitstream/handle/20.500.11822/9530/-Climate_Change_and_Human_Rightshuman-rights-climate-change.pdf.pdf?sequence=2&amp%3BisAllowed= (accessed August 2020).

UNWTO (United Nations World Trade Organisation) (2020) International tourism highlights 2019. See https://www.e-unwto.org/doi/pdf/10.18111/9789284421152 (accessed January 2020).

Yeoman, I. and McMahon-Beattie, U. (2019) The experience economy: Micro trends. *Journal of Tourism Futures* 5 (2), 114–119, doi:10.1108/JTF-05-2019-0042.

Yeoman, I. and McMahon-Beattie, U. (2020) Introduction: Does the past shape the future? In I. Yeoman and U. McMahon-Beattie (eds) *The Future Past of Tourism* (pp. 1–8). Bristol: Channel View Publications.

Index

Accessibility xi, 59, 61, 65, 67, 69–70, 76, 107–117, 150
Accessibility information 113–114, 116–117
Accessible hospitality xi, 61, 70
Accessible tourism xi, 2, 15, 59, 61–63, 107–108, 110–112, 114–115
Acting 8, 126, 132, 135, 138
Actors 1–7, 16, 19, 23, 67, 72–73, 86, 90–91, 144–145, 147, 150–152
Aesthetic capital 7, 35, 40, 42, 49
Aesthetic labour 35, 41–42, 44, 49–53
Airbnb 7, 33–39, 41–52
Airbnb community 36, 43, 45–48
Airbnb host 35, 39, 43, 47, 49, 51
Algorithm 36, 39–40, 42, 44, 48, 50, 53
Alternative futures 4, 7, 60, 65, 77, 150, 152

Barrier-free tourism 61, 111
Belonging 1, 8, 62, 126, 132, 135, 138
Benefit 18, 67, 82
Brazil 9, 80–81, 87–89, 93–95, 99

Causal Layered Analysis (CLA) 7, 60, 65–67, 75
Cooperatives 127–128
Community-based tourism 2, 17, 25, 82
Competences 34–35, 37, 49, 52–53
Correctness and completeness 114
COVID-19 4, 144–145, 147, 149–150
Credibility 50–51, 91, 113, 115
Crowdsourcing 117
Cultural capital 42, 51, 53, 61
Cultural sensitivity 16–17, 22–24, 27

Destination apps 113
Destination websites 113

Disabled 1, 111, 134, 140
Disabled tourism 111
Down-syndrome 123

Easy-access tourism 111
Elderly tourism 111
Emotional capital 47–49
Emotional labour 41, 49, 51
Employment mismatch 132–133
Empowerment 9, 18–19, 22
Engagement 2, 4, 26–27
Ethnic minorities 1–2
Exclusion 45, 63, 83, 125–127

Finland 18–22, 43–44, 60–68, 72, 75, 122–135, 139–140, 150
Future 2–9, 53, 60, 64–67, 71–76, 97, 110, 117, 125, 134, 139–140, 144–152

Government programme 134, 139–140
Guest 17, 20–21, 47–48, 50–51, 59, 94, 151

Having 126, 132–135, 138
Highlands and Islands 64–65, 68–72
Hospitality xi, 9, 16, 20–21, 33–36, 39–42, 47–48, 51, 53, 62, 67, 73–75, 123, 149
Host 17, 20–22, 38–39, 41, 43–49, 59, 61, 75
Host-guest relations 16–17, 20, 26, 66–67, 74–75

Impact/s 3–5, 17, 37, 67, 69, 82, 86, 90, 94, 97, 99, 136, 147–148, 152
Inclusion 1–3, 6, 9, 15–16, 19, 22–24, 26, 52, 59–63, 66–67, 69–70,

74–76, 125–126, 128, 132–133, 135, 137–140, 144–150
Inclusive tourism 1–6, 9, 15, 25, 59–60, 66, 82, 111, 144, 146–147, 149–152
Income 36, 81–84, 86–89, 92, 96–100, 134, 136–137
Indigenous people 2, 18–21, 24, 148
Indigenous tourism 17–18, 21–22, 24
Inequality 33, 36–38, 41, 52–53, 82–83, 89, 99–100, 147, 150
Information content 109
Inner possibilities 62
Interview 23, 124–125, 139, 141

Labour demand and supply 131, 140
Lapland xi, 9, 64–65, 68, 70, 73–74, 124–125, 128–133, 135–141, 150
Layer 60, 65–67, 74
Litany 65–69, 71, 75
Local 1, 5, 18–20, 25, 69–73, 86, 88–90, 92–94, 96–97, 117, 135–139, 147, 150
Local participation 2, 15, 17, 19, 24, 26, 69, 76
Local communities 15, 18–19, 24–26, 73, 81–82, 86, 88, 97, 111, 138

Madeira 64–65, 68–69, 71–73, 75
Meritocracy 35, 37–39, 53
Metaphors 59–60, 63, 74
Methods 3, 6, 8, 18, 60, 64, 100, 144, 147, 151
Myth 35, 67, 74–76

Online hospitality 35, 40, 45, 53
Online performance 34, 41
Online platforms 108–109, 113, 147
Openness 20, 53, 62, 73
Opportunity Study Guidelines 86–87, 91–92
Outer possibilities 62, 66–67, 73
Overall inclusion 109

Participatory approach 15, 25, 67, 84, 90, 100, 147
Participatory tourism 2, 15–16, 18–19, 23–26
Participatory tourism development 15, 18
Participatory tourism projects 16, 18, 25–26

People with disabilities 107–109, 115–117, 123–124, 150
Planning 4, 9, 17, 19, 23–26, 59–60, 63–64, 69, 82, 85, 108–110, 148
Platform economy 33–37, 40–42, 47, 49, 52–53
Platform work 35, 52
Portugal 64–65, 70, 75
Post-structuralism 62, 66
Poverty 61, 80–84, 86, 88–89, 98–100, 126
Poverty line 84, 89, 96, 98,
Poverty reduction 80–82, 86
Practices 3, 6, 15, 34, 41–42, 67, 113, 140, 144, 148, 151–152
Pre-trip stage 108–109
Project 15–27, 64, 128, 132
Project community 26
Project partners 23–24, 27
Project planning 23–24
Project society 16, 27
Pro-poor tourism (PPT) 2, 80–82, 84, 92, 97–98
Pro-poor VCA 80–81, 85–87, 89, 92–93, 97–100
Public policy 111

Qualitative 43, 84, 100, 116
Quantitative 66, 69, 84, 100, 116

Rainbow tourists 1
Refugees 124
Regional development plan 140
Reliable information 115–116
Resources for participation 19, 23–27, 33–35, 40–42, 52–53, 77, 83, 86, 108–109, 126, 132, 149
Revenue 42, 70–72, 86–87, 92–97, 99, 135

Sámi tourism 18, 21–22, 137
Scotland 60, 64–65, 68–69, 71–72
Seasonality 131, 133–138
Sensitive tourism 16, 22–24
Service industries 131
Sharing economy 34, 47–49
Smart destination 112
Smart specialization platform 140
Social economy 123–125, 127–128, 140
Social enterprise 127–128, 151
Social entrepreneurship 5, 15, 127–128, 148

Social inclusion 9, 123–128, 132, 139–141
Social inequality 36, 38, 41, 125
Social sustainability 9, 80, 82, 99–100, 123, 149
Social tourism 1–2, 5, 15
Staged participation 75–77
Stakeholders 1–3, 16, 19–20, 22–27, 70–73, 76, 90–91, 111, 138–139, 144, 147, 150–151
Stratification 34–36, 40, 43, 48, 53
Sustainability 4, 69, 71, 76, 112, 122–124, 137, 141, 145, 149–150
Sustainable development 17, 88, 98–100, 122, 140, 150
Sustainable development goals 80, 122–123
Symbolic distinction 35
Systemic 40, 42, 67, 69–71, 76

Talent 37–38, 45–46, 53
Tourism accessibility standards 109
Tourism for all 2, 61, 111, 114, 152
Tourism planning 17, 25, 59, 69, 82
Tourism strategy 59–60, 64–66, 68–71, 73–75, 77, 88, 134–135, 138, 140
Tourism value chain 80, 85, 90–93, 95–97
Temporary communities 26

Universal accessibility 107, 110–112
Updated information 109
Updatedness 114–115
User perceptions 116

Value 9, 20, 36, 41–42, 52, 59–60, 69, 73, 81, 128, 150–151
Visibility (of information) 114–115
Vocational education and training (VET) 125, 133–135

Web accessibility 109
Welfare state 43, 127
Well-being 15–16, 18, 20, 24–25, 59, 74, 84, 89, 98, 123, 126, 145, 150
Western world 5, 64, 75, 149
Workforce 125, 131–133, 139, 141
Worldview 5, 19, 60, 63, 67, 71, 73–74, 76–77, 149

Youth 147–148

For Product Safety Concerns and Information please contact our EU Authorised Representative:

Easy Access System Europe

Mustamäe tee 50

10621 Tallinn

Estonia

gpsr.requests@easproject.com

www.ingramcontent.com/pod-product-compliance
Ingram Content Group UK Ltd.
Pitfield, Milton Keynes, MK11 3LW, UK
UKHW021943200326

4879IPUK00004B/64